Nature's Machines

To Helen, Diane, and Kevin
My favorite traveling companions!

Nature's Machines
An Introduction to Organismal Biomechanics

by

David E. Alexander
Dept. of Ecology & Evolutionary Biology
University of Kansas, Lawrence, KS, United States

ACADEMIC PRESS
An imprint of Elsevier

Academic Press is an imprint of Elsevier
125 London Wall, London EC2Y 5AS, United Kingdom
525 B Street, Suite 1800, San Diego, CA 92101-4495, United States
50 Hampshire Street, 5th Floor, Cambridge, MA 02139, United States
The Boulevard, Langford Lane, Kidlington, Oxford OX5 1GB, United Kingdom

Notices
Knowledge and best practice in this field are constantly changing. As new research and
experience broaden our understanding, changes in research methods, professional practices,
or medical treatment may become necessary.

Practitioners and researchers must always rely on their own experience and knowledge in
evaluating and using any information, methods, compounds, or experiments described
herein. In using such information or methods they should be mindful of their own safety and
the safety of others, including parties for whom they have a professional responsibility.

To the fullest extent of the law, neither the Publisher nor the authors, contributors, or editors,
assume any liability for any injury and/or damage to persons or property as a matter of
products liability, negligence or otherwise, or from any use or operation of any methods,
products, instructions, or ideas contained in the material herein.

Library of Congress Cataloging-in-Publication Data
A catalog record for this book is available from the Library of Congress

British Library Cataloguing-in-Publication Data
A catalogue record for this book is available from the British Library

ISBN: 978-0-12-804404-9

For information on all Academic press publications visit our website at
https://www.elsevier.com/books-and-journals

Working together
to grow libraries in
developing countries

www.elsevier.com • www.bookaid.org

Publisher: Sara Tenney
Acquisition Editor: Kristi Gomez
Editorial Project Manager: Pat Gonzalez
Production Project Manager: Julia Haynes
Designer: Matt Limbert

Typeset by TNQ Books and Journals

Contents

Preface   ix

1.   Introduction and Physics Review

   1.1   What Is Biomechanics?   1
   1.2   A Brief History of Organismal Biomechanics   3
   1.3   Review of Newtonian Physics   5
         1.3.1   Vectors and Scalars   5
         1.3.2   Newtonian Mechanics and the Work—Energy Relationship   5
         1.3.3   Newton's Second Law: Force and Movements   7
         1.3.4   Unsteady Motion   9
         1.3.5   Derived Quantities Involving Force   9
         1.3.6   Mass and Weight   10
         1.3.7   Other Physical Quantities   13
         1.3.8   Units and the SI System   13
         Further Reading   13
         General References   13
         Locomotion   13
         Material Properties   14
         Sports Biomechanics   14

2.   Solid Materials

   2.1   Introduction to Solids   15
   2.2   Loading, Deformation, Stress, and Strain   16
         2.2.1   Loads and Deformations   16
         2.2.2   Stress and Strain   17
         2.2.3   Information From Stress—Strain Curves   19
         2.2.4   Nonlinearity   23
         2.2.5   Strength Versus Toughness   26
         2.2.6   Biological Examples   28
   2.3   Failure and How to Prevent It   30
         2.3.1   Fracture Mechanics: All About Cracks   32
         2.3.2   How to Stop Crack Growth   35
   2.4   Structures   38
         2.4.1   The Engineering Categories   38
         2.4.2   Elongate Structures: Beams   39
         2.4.3   Elongate Structures: Columns   42
         2.4.4   Shells   44

2.4.5 Examples of Biological Structures   45
Further Reading   49
General References, Classic   49
General References, Recent   50
Collagen and Tendons   50
Plant Mechanics   50
Protein Rubbers   50
Shell and Bone   50
Silk   50

3. Fluid Biomechanics

3.1 **Fluid Basics**   51
   3.1.1 Fluids Defined and How to View Them   51
   3.1.2 Viscosity   53
   3.1.3 Drag   54
   3.1.4 The Reynolds Number   57
   3.1.5 Drag Reduction in Swimmers and Flyers 60
   3.1.6 Drag as an Asset   62
3.2 **Fundamental Equations**   63
   3.2.1 Bernoulli's Equation   63
   3.2.2 The Navier–Stokes Equations   64
3.3 **Velocity Gradients and Boundary Layers**   65
   3.3.1 What Is a Boundary Layer?   65
   3.3.2 Boundary Layer Thickness   65
   3.3.3 Living in Boundary Layers   66
3.4 **Wings and Lift**   69
   3.4.1 The Lift Mechanism and the Bound Vortex 69
   3.4.2 Modifying Lift   72
   3.4.3 Induced Drag Causes and Consequences 74
   3.4.4 Gliding and Flapping   77
   3.4.5 Hydrofoils   81
   3.4.6 Wings and Size   82
3.5 **Swimming**   84
   3.5.1 Swimming Modes   85
   3.5.2 Lift-Based Swimming   85
   3.5.3 Drag-Based Swimming   85
   3.5.4 Undulatory Swimming   87
   3.5.5 Swimming by Jetting   89
   3.5.6 Swimming at Low Reynolds Numbers   89
3.6 **Internal Flows**   90
3.7 **When Flows Are Not Steady**   93
   3.7.1 Continuous Acceleration: The Acceleration Reaction 93
   3.7.2 Unsteady Effects in Air   95
   Further Reading   96
   General   96
   Flight   97
   Internal Flows   97
   Swimming   97
   Unsteady Processes   97

4. Biological Materials Blur Boundaries

4.1	Viscoelastic Solids	99
	4.1.1 Transient Tests	99
	4.1.2 Springs and Dashpots	101
	4.1.3 Dynamic Testing	102
	4.1.4 Biological Examples of Viscoelastic Materials	105
4.2	Non-Newtonian Liquids	111
	4.2.1 Non-Newtonian Behavior of Everyday Liquids	112
	4.2.2 Blood	112
	4.2.3 Synovial Fluid	113
	4.2.4 Biological Liquids in General	114
4.3	Mucus	114
	4.3.1 Snail Pedal Mucus	115
	4.3.2 How Slugs Glide	116
4.4	Swimming in Sand: Locomotion in Granular Media	117
	Further Reading	119
	General	119
	Locomotion in Granular Media	119
	Non-Newtonian Fluids	119
	Mucus	120
	Viscoelastic Materials	120

5. Systems and Scaling

5.1	Putting It All Together: Biomechanics in Action	121
5.2	Legs: Muscles, Joints, and Locomotion	121
	5.2.1 Muscle Biomechanics and Scaling	121
	5.2.2 Articulations: Adding Flexibility to Rigid Skeletons	124
	5.2.3 Locomotion on Two (or More) Legs	126
5.3	"Soft" (Hydrostatic) Skeletons	132
	5.3.1 Muscles and Stresses in the Wall	133
	5.3.2 Fiber-Reinforced Hydrostats	133
	5.3.3 Biological Examples	136
	5.3.4 Muscular Hydrostats	137
5.4	The Consequences of Size	138
	5.4.1 Surface to Volume Ratio	139
	5.4.2 Maximum Jump Heights	141
	5.4.3 Growing Into Different Mechanical Realms?	143
5.5	The Promise of Biomimicry: Have We Arrived?	146
	5.5.1 Ornithopters	147
	5.5.2 Adhesives	147
	5.5.3 Legged Robots	148
	Further Reading	149
	Biomimicry	149
	Hydrostatic Skeletons	150
	Maximum Jump Heights	150
	Muscles and Locomotion	150
	Size and Scaling	150
	Terrestrial (Legged) Locomotion	150

6.   Organismal Versus Technological Design

6.1	Borrowing From Engineers	151
6.2	Different Materials Used in Different Ways	152
	6.2.1 Materials	152
	6.2.2 Shape	152
	6.2.3 Loading and Movement	153
	6.2.4 The Construction Process	154
6.3	Research and Methods	155
	Further Reading	156

Bibliography	157
Index	171

Preface

This book grew out of conversations with colleagues who saw a need for a concise, introductory-level primer on organismal biomechanics. A number of introductory books that cover human-oriented biomechanics are available, but the only book with broad coverage of biomechanics of organisms other than humans is Steven Vogel's *Comparative Biomechanics* (2013). Vogel's book is authoritative, witty, insightful, and will probably be the gold standard against which future general biomechanics books will be judged. At over 600 pages, it is more advanced and has broader and deeper coverage than might be desired in an entry-level book. Nevertheless, his description of organismal (comparative) biomechanics would be difficult to improve upon: "Comparative biomechanics...starts with the ordinary activities of ordinary organisms, posing questions a person might ask while exploring a coastline or tramping through a forest...Comparative biomechanics invokes notions closer to one's everyday experience and intuitive sense of reality than does any other area of contemporary science. Gravity and elasticity have an immediacy that cells and molecules, let alone galaxies and subatomic particles, do not."

My aim is for this book to be accessible to a typical science undergraduate, or anyone who has had some high school or college physics. Biomechanics is essentially applied Newtonian physics, so at least some familiarity with basic concepts—Newton's three laws of motion, work, energy, and power—is necessary. I have provided a physics review in Chapter 1 that covers the basic physics needed, although to really understand such concepts, working through problem sets on mechanics from any introductory physics textbook might be useful for those who have not had a formal course. I hope my colleagues working in other areas of biology also might find this book useful, either as an introduction to the biomechanics research literature or as a resource when their own research might benefit from a biomechanics perspective.

In this book, my goal was to cover most of the fundamental concepts likely to be encountered in studies of animal and plant biomechanics. An author's choices cannot help but be influenced by his or her interests and expertise; I have thus explicitly limited coverage to macroscopic organisms, and the biomechanics of locomotion probably received more than its fair share of coverage. In striving to achieve a concise presentation, I gave some of the more sophisticated topics only superficial coverage and omitted a few entirely. I have cited sources extensively, both important reference books and research articles, so the interested reader should be able to pursue any topic in this book in as much depth as desired.

As others have noted, biomechanics draws on diverse fields whose practitioners may not interact much with each other and have developed their own conventions for symbols (e.g., dynamic viscosity is μ (Greek mu) in some fields and η (Greek eta) in others). I have tried to point out where more than one symbol for a quantity is in common use and specify which (if either) is more common in biomechanics usage. Conversely, different fields may use the same symbol for different things, such as using τ (Greek tau) for both torque and the relaxation time for a viscoelastic material or δ (Greek delta) for boundary layer thickness and the viscoelastic phase angle. I have defined such quantities where they are used, and fortunately for our purposes, the relevant quantity is almost always obvious in context. Readers should not assume, however, that symbols will always be used in the research literature as I have defined them in this book.

I was coincidentally asked to write a review of general biomechanics by the *European Journal of Physics* just before I began work on this book (Alexander, 2016). The research for that review gave me a very useful overview of the field, and writing it helped me develop a framework for organizing the material. The review article also acted as a sort of backbone for Chapters 2 and 3, and parts of Chapters 5 and 6. I was happy to be able to include some material in this book that I was unable to include in the review article (which nevertheless may be the longest review article ever published by *EJP*).

Given the nature of this book, pretty much anyone I have interacted with professionally over the years has had some positive impact on the book. They are literally too numerous to mention, as much as I wish it were otherwise. The late Steven Vogel, my graduate advisor and long-time mentor, has always been my main inspiration; I hope this book at least approaches his high standards. Jeff Dawson first got me thinking about the idea of a biomechanics primer, and Robert Dudley enthusiastically endorsed the idea. The students in my seminar classes (animal locomotion, biomechanics, and animal flight) over the years gave me much practice in introducing basic biomechanics to undergraduates. My discussions with the participants of various biomechanics-related symposia at the XXV International Congress of Entomology, especially John (Jake) Socha, David Lentink, Catherine (Kate) Loudon, Sanjay Sane, Günther Pass, and Yoshinori Tomoyasu, were very fruitful and encouraging. I greatly appreciate the patience of Sara L. Taliaferro as she dealt with the conversion of my sometimes vague concepts into crisp illustrations. She drew most of the figures in the book, as indicated in the relevant figure legends. (Fig. 3.15 was originally drawn by Barbara Heyford for my first book, and I drew all of the remaining figures.) Rémy Lequesne provided me with cast iron stress−strain curves collected by his students and found reliable Young's modulus values for glass. We also had a helpful discussion about usage of "Young's modulus" and standard engineering terminology. Ken Fischer gave me an engineer's perspective on properties of organic materials. Tom Daniel helped me reconcile what at first appeared to be multiple inconsistent ways of looking at the mechanics of undulatory swimming. Mark Denny, Roland Ennos, and Kate Loudon each read sections of the manuscript and gave me thoughtful, helpful suggestions (and pulled me back from assorted blunders); along with two anonymous reviewers of the original book proposal, they greatly improved the final product. Any errors that remain are, of course, my own responsibility. Finally, I must thank my wife and best friend, Helen Alexander, for her unflagging support and ever-present encouragement.

Chapter 1

Introduction and Physics Review

1.1 WHAT IS BIOMECHANICS?

A dictionary might define biomechanics as "the mechanics of biological activity" (Mish, 1983). Most scientists who actually perform biomechanics research tend to describe biomechanics as something like engineering in reverse: engineers are given a task and they design a device to carry out the task, whereas biomechanics researchers have the "device" (an organism) and they seek to understand its mechanical properties and constraints, and how it operates. Nowadays, at least for nonhuman organisms, this is usually put in an evolutionary framework—what are the benefits that might have selected for a given mechanical arrangement, and how does that arrangement compare to those of the organism's relatives? Traditionally, biomechanists have said that they take engineering approaches and techniques and apply them to help understand how organisms work. In recent decades, many biomechanics researchers have actually moved beyond standard engineering approaches and developed new methods better suited to biological subjects, as we will see in later chapters.

Under the subject listing of "biomechanics," the catalog of a typical university library will group together books in several areas. One of these areas will be clinical or medical biomechanics; the wonderfully concise book *Introduction to Biomechanics* by Frost (1967) in this area was part of the inspiration for this book. Clinical biomechanics focuses on humans, only involving nonhumans to the extent that they help understand human mechanics. This area was originally developed by and for surgeons, who needed to know how their surgical interventions affected a patient's ability to function mechanically. As artificial joints and other musculoskeletal repairs and implants came into use, the focus of clinical biomechanics shifted toward orthopedic medicine, as well as analyzing normal body structure and movements, and the mechanical consequences of pathological conditions.

Clinical biomechanics overlaps considerably with biomedical engineering; many books on the latter will also be found under the "biomechanics" subject heading. Biomedical engineers design medical equipment, including

Nature's Machines. http://dx.doi.org/10.1016/B978-0-12-804404-9.00001-3

1

prosthetics, implantable devices, surgical equipment and processes, ultrasonic and other imaging systems, and devices for treatment and rehabilitation such as dialysis machines and physical therapy equipment, among many other things. Biomedical engineers also perform the same sorts of analyses on body structure and movement as do clinical biomechanics researchers, but with a more engineering-oriented perspective. Biomedical engineers and clinical biomechanists often form part of a research team, bringing a variety of approaches to a research problem.

Another large cluster of books under the "biomechanics" heading in the library catalog consists of those aimed at sports, athletic performance and training, and general human fitness. Sports biomechanics is largely focused on improving athletic performance or reducing sports injuries. For example, research on the mechanics of running led to improvements in both running shoes and resilient running tracks. Part of the goal was to try to increase running speeds, but what was originally a side benefit of reducing running-related injuries has now become a major priority. Sports biomechanics has focused heavily on running and swimming, but it has also produced some fascinating results in areas such as the sensory tasks involved in batting and catching high fly balls in baseball (e.g., Cross, 2011; Higuchi et al., 2016).

Taken together, these three human-oriented areas—clinical biomechanics, sports biomechanics, and biomedical engineering—make up *applied biomechanics*. Each area has its own specialist practitioners and its own types of problems, but they overlap a fair amount. Sports biomechanics encompasses some of the same topics as clinical biomechanics, especially in the area of normal locomotion mechanics. Understanding the mechanics of human locomotion may be just as important for researchers designing prosthetic limbs as for those working on improved athletic shoes. Indeed, sports and clinical biomechanics merge in physical therapy and rehabilitation research.

Perhaps the smallest cluster of "biomechanics" books in the university library consists of those books on the biomechanics of nonhuman organisms. This discipline was originally called "comparative biomechanics" to distinguish it from applied biomechanics, just as "comparative physiology" traditionally referred to physiology of nonhuman animals. More recently, some systematists[a] co-opted "comparative" as used in the phrase "the comparative method" to describe a particular phylogenetic approach to biological research (Felsenstein, 1985), quite unrelated to the traditional "comparative" versus "applied" meaning. To avoid any confusion, I prefer the term *organismal biomechanics*. As generally understood by scientists, comparative or organismal biomechanics focuses on macroscopic organisms, essentially macroscopic animals and plants. Certainly microscopic organisms experience

a. Systematists are biologists who study the patterns of evolutionary relationships among organisms.

mechanical processes and possess mechanical properties, but those processes and properties are so different from the mechanics of macroscopic organisms that they represent a completely different subset of mechanics. So this book will focus on the mechanics of macroscopic animals and plants.

The dictionary definition of "biophysics" is "the application of physical principles and methods to biological problems" (Mish, 1983). In spite of this definition, biomechanics in practice is not a subdiscipline of biophysics. Research in the field of biophysics, as it is actually performed and reported in its premier journals, consists mainly of studies at the subcellular and molecular level. Some typical topics in biophysics research include molecular mechanisms in muscles and other intracellular movement mechanisms, molecular biology of membrane potentials and other cell membrane processes, and protein structure and function. The early practitioners of this research coined the term "biophysics," and by convention, it has retained its concentration on subcellular processes to the present time. In contrast, the term "biomechanics" was coined decades later; from the start it was focused on the mechanics of humans (Zarek, 1959) and macroscopic organisms (Brown, 1948; Osborne, 1951) and has generally focused on material properties of solid structures, fluid mechanics of environmental flows or body fluids, and the kinematics and dynamics of organisms or their parts.

1.2 A BRIEF HISTORY OF ORGANISMAL BIOMECHANICS

Long before the term "biomechanics" (or even "biology") was coined, many of the early natural philosophers, physicists, and physiologists performed studies that we would today classify as biomechanics. As recounted by Y. C. Fung (1981) in his book, *Biomechanics: Mechanical Properties of Living Tissues*, Galileo and many of his successors—Giovanni Borrelli, Robert Hooke (Hooke's law), Robert Young (Young's modulus), Jean Poiseuille (Poiseuille's equation), Hermann von Helmholtz (the law of conservation of energy)—may have been better known in other areas, but all made studies in areas we would now recognize as biomechanics or biomechanics-related physiology.

Although modern engineers or mechanically inclined biologists made the occasional biomechanical study in the first half of the 20th century, biomechanics seems to have first come into its own as a recognized research field in the 1950s and 1960s. Many organismal biomechanics research studies from that period focused on animal locomotion, particularly flight. Some of the earliest examples are Brown's work on bird flight mechanics (Brown, 1948, 1953) and that of Osborne (1951) on insect flight. Soon after, a pioneering biologist—engineer collaboration led to Torkel Weis-Fogh and Martin Jensen's classic series of papers on the mechanics of locust flight (Jensen, 1956; Weis-Fogh, 1956; Weis-Fogh and Jensen, 1956). At about that time, other researchers began to measure the mechanical properties of biological materials, e.g., the springy hinge ligament of scallops (Trueman, 1953), insect mouthparts (Bailey, 1954), sisal leaf fibers (Balashov et al., 1957), the walls of fish swim bladders (Alexander, 1959a,b), sea anemone body wall (Alexander, 1962), rat uterus (Harkness and Harkness, 1959), and rat tail tendon (Rigby et al., 1959). Biologists interested in biomechanics began to adopt a quantitative, engineering-inspired approach to topics such as the strength of biological materials, particularly in relation to biologically relevant environmental loads and mechanics of body support (e.g., Alexander, 1962). In addition to swimming (Bainbridge, 1958) and flying (Boettiger and Furshpan, 1952), biologists interested in fluid mechanics also began studying topics such as fluid mechanics of suspension feeding (Jørgensen, 1955).

Biomechanics experienced its first great expansion in the 1960s and 1970s. A number of pioneering researchers began their research careers just as biomechanics emerged as a recognized discipline, and they helped define the field. Some of the most influential researchers were R. McNeill Alexander, Julian Vincent, and Stephen Wainwright—who mainly specialized in solid mechanics—and Werner Nachtigall, Ulla Norberg, Colin Pennycuick, and Jeremy Rayner—animal flight specialists who focused heavily on aerodynamics. Stephen Vogel started out studying primarily fluid biomechanics but eventually became something of a biomechanics polymath, working on fluid–solid interactions and properties of biological materials. Most of these pioneers wrote books that became the standard works in their fields, e.g., Alexander (1968), Pennycuick (1972), Wainwright et al. (1976), Vogel (1981), and Vincent (1982). Although these books obviously do not cover the most recent research in biomechanics, they still provide useful introductions to their respective topics, and most are still valuable references to this day.[b]

In the 1980s and 1990s, many students of this pioneer group of researchers went on to expand and extend biomechanics beyond its engineering-inspired roots. A few examples out of a huge number of possibilities include Charles Ellington, Robert Dudley, Kenneth Dial, and Sharon Swartz (flight biomechanics); Thomas Daniel (swimming biomechanics); William Kier (muscle and soft tissue mechanics); Andrew Beiwener (bone and muscle mechanics); and Roland Ennos, Mimi Koehl, and Mark Denny (both fluid and tissue biomechanics). These researchers have increased the visibility and relevance of biomechanics among biologists while maintaining a level of rigor that compares favorably with biomedical engineering. A brief sample of classic studies from this period includes studies of flight in insects (Ellington, 1984b; Ennos, 1988), birds (Spedding, 1986), and bats (Swartz et al., 1992); mechanics of unsteady swimming (Daniel, 1984);

---

b. Given its status as a classic work in the field, readers should be aware that Wainright et al. (1982) is not a second or revised edition of Wainright et al. (1976) but essentially a reprint by a different publisher. The same applies to Vincent (1982, 1990). In contrast, Vogel (1994) is a revised and expanded second edition of Vogel (1981); Vogel's first edition, however, contains a useful appendix of practical methods that was omitted from the second edition.

and mechanical properties of soft tissue (Kier and Smith, 1985), spider silk (Denny, 1980), and snail mucus (Denny, 1984).

Recent decades have seen biomechanics expand into new organisms and new areas, e.g., pterosaur wing mechanics (Bennett, 2000), fractures in tree branches (Ennos and van Casteren, 2010), and force generation in climbing plants (Isnard et al., 2009). The complexity and sophistication of biomechanical analyses continue to increase, e.g., finite element analyses (Herbert et al., 2000; Manning et al., 2009) and computational fluid dynamics (Liu and Sun, 2008). Moreover, biomechanical analyses have come to be expected in areas such as vertebrate paleontology, locomotion physiology, and feeding biology.

1.3 REVIEW OF NEWTONIAN PHYSICS

This book assumes that the reader has already received an introduction to basic physics. The next section is intended to serve as a brief refresher of key Newtonian concepts and relationships. For those who have not had such an introduction, I recommend first reading the mechanics section of any introductory physics textbook (e.g., Halliday et al., 2010) or one of the problem-based guides such as the Schaum's Outline Series (Nelson et al., 1997).

1.3.1 Vectors and Scalars

The physical quantities used in this book will all be either vectors or scalars. Scalars are quantities with only magnitude, such as energy or temperature. Vectors are quantities with both magnitude and direction, such as force or acceleration. Vectors are typically written in bold (**a**) or with an overprinted dash or arrow ($\overline{a}$, $\overrightarrow{a}$) to distinguish them from scalars. In many calculations we only need the magnitude of the vector, so we can use the symbol without bold or overprinting to represent the magnitude alone. Most of the equations with vectors in this book will be arranged in such a way that vectors are parallel or perpendicular so that calculations can be performed as if they were scalars, using only the magnitude. Nevertheless, when situations involve forces, velocities, or accelerations, we must keep in mind that direction in the real world may not be so conveniently organized and may require explicit vector algebra. Finally, by convention, *velocity* is a vector, whereas *speed* is the magnitude of the velocity and is thus a scalar. This convention can be confusing when "v" is used as a symbol for the speed (magnitude) of the velocity "**v**."

1.3.2 Newtonian Mechanics and the Work–Energy Relationship

Biomechanics is an application of Newtonian mechanics. Classical Newtonian mechanics is the physics of forces and movements at scales larger than atoms.

BOX 1.1 Newton's Three Laws of Motion

1. Newton's first law ("law of inertia"): An object at rest will stay at rest, and an object in uniform motion will stay in motion, unless acted upon by an outside force.
2. Newton's second law: Force equals mass times acceleration ($\mathbf{F} = m\mathbf{a}$). This can be restated (in a form closer to Newton's original) as the acceleration of an object is proportional to and parallel to the force applied to the object and inversely proportional to the mass of the object.
3. Newton's third law ("law of equal and opposite reactions"): For every object exerting a force on a second object, the second object exerts a force equal in magnitude and opposite in direction.

It incorporates kinematics, focused on motion, and dynamics, focused on forces. This is generally the physics of our everyday experience, governed by Newton's three laws of motion (see Box 1.1). Mass cannot be created or destroyed, and energy is constant in closed systems. We do not need to consider relativity, quantum mechanics, or even compressibility (high Mach effects) or electricity and magnetism.

The fundamental scalars in mechanics are work and energy. Work is done when some force causes an object to move. The amount of work is the magnitude of the force integrated over the distance moved; for a constant force, work (W) is simply the magnitude of the force (F) times the distance (d), $W = Fd$. Energy is defined as the ability to do work, which simply means that energy is anything that can be used to generate force times distance. In mechanics, energy is typically in the form of kinetic energy ($\frac{1}{2} mv^2$)—energy of speed—or gravitational potential energy (mgh)—the energy of position (where g is the acceleration due to gravity, h is the height above some reference level, m is the mass, and v is the speed). Many other forms of energy exist—heat, voltage, chemical bond energy—but these forms play little direct role in mechanics. Both work and energy have unit of joules (J) in the standard SI system of units.

The law of conservation of energy is extremely important in the derivation of many fundamental mechanical principles and relationships. This law is less important in practical and empirical research for at least two reasons. Even in a closed system, certain forces such as friction generate "nonconservative" work, which produces an increase in internal energy (basically heat). Internal energy tends to dissipate and is lost as a mechanically useful quantity. Moreover, closed systems in the physics sense are rare in biology, so while it is true that energy cannot be created or destroyed in everyday situations, if thermal or other energy is free to enter or leave the system, keeping track of all the work and energy can become very unwieldy.

1.3.3 Newton's Second Law: Force and Movements

Velocity, **v**, may be the most intuitively simple vector. Its magnitude, speed (v), is the distance moved per unit time. For a constant speed, this is just the total change in displacement (Δx) divided by the change in time (Δt), v = Δx/Δt. For speeds that are not constant, speed becomes v = dx/dt. Speed is thus a rate, i.e., the time derivative of displacement. The SI system has no special unit for speed, the units are just used in their fundamental form of meters per second (m s^{-1}).

Acceleration, **a**, is the rate of change of velocity. If the velocity's magnitude *or* its direction (or both) changes, that change can be represented by an acceleration. The acceleration's direction need not have any relationship to the velocity's direction. In the simplest cases, the acceleration is parallel to the velocity, and either in the same direction—speed increases—or in the opposite direction—speed decreases—with no change in the velocity's direction (Fig. 1.1A and B). If the acceleration is not parallel with the velocity, the object will follow a curved path. If the acceleration is perpendicular to the velocity (Fig. 1.1C), the object will travel in a circle, where the velocity constantly changes its direction while maintaining a constant speed (magnitude). Accelerations, like velocities, have no special units in the SI system; they are simply given as meters per second squared (m s^{-2}).

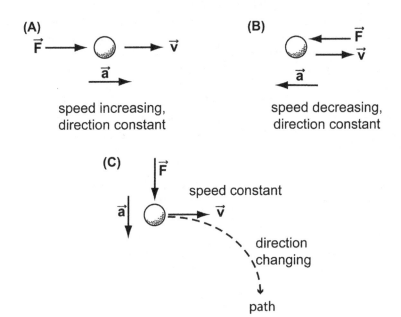

FIGURE 1.1 Force, speed, and acceleration of a particle. (A) Force and speed parallel and in the same direction. (B) Force and speed parallel and in the opposite direction. (C) Force and speed not parallel (in this case at right angles, giving a circular path). *Artist: Sara Taliaferro.*

In everyday terms, a force is a push or pull. In physics, force is defined by Newton's second law: $\mathbf{F} = \mathbf{ma}$ (see Box 1.1). A force is thus any push or pull that could potentially produce an acceleration, where the magnitude of the force is proportional to the magnitude of the acceleration and the direction of the force is in the same direction as the acceleration. Rearranging Newton's second law to read $\mathbf{F}/m = \mathbf{a}$ emphasizes that the acceleration will be inversely proportional to the mass of the object being accelerated. Mass in the SI system is a fundamental quantity with the unit kilogram (kg). Force is a derived unit in the SI system, where its aptly named unit, the newton (N), is equivalent to $kg\ m\ s^{-2}$ in fundamental units.

Momentum is a concept closely intertwined with force. *Linear momentum*, **P**, is defined as mass times velocity, mv. A simple substitution shows that the rate of change of momentum equals force:

$$\frac{d\vec{\mathbf{P}}}{dt} = m\vec{\mathbf{a}} = \vec{\mathbf{F}} \tag{1.1}$$

In fact, Newton actually stated his second law in terms of the rate of change of momentum rather than mass and acceleration. Under the right conditions, momentum, like energy, is conserved. In a so-called elastic collision, and neglecting friction, the total momentum of bodies in a collision will be the same before and after the collision. This allows speeds and directions to be predicted after objects collide. Few biological materials, however, produce elastic collisions, so such predictions are rarely accurate for biological collisions. Momentum does not have its own SI unit, so it is simply expressed in terms of fundamental units, $kg\ m\ s^{-1}$, which is often simplified to N s. Rate of change of momentum, however, has the same unit as force, i.e., newton (N).

Rotational motion follows essentially the same relationships as linear motion, with angular velocity, ω, replacing linear velocity, **v**. Thus, angular acceleration, α, is the rate of change of angular velocity. Torques or moments are the angular analogs of forces. Torque, τ, is given by

$$\vec{\tau} = \vec{\mathbf{F}} \times \vec{\mathbf{r}} \tag{1.2}$$

where **F** is a force and **r** is the position vector from the center of rotation to the point of application of the force. Rotational inertia, l, is

$$\vec{\mathbf{l}} = \vec{\mathbf{r}} \times \vec{\mathbf{P}} \tag{1.3}$$

where **P** is linear momentum, and by analogy with linear momentum, torque equals the rate of change of angular momentum:

$$\vec{\tau} = \frac{d\vec{\mathbf{l}}}{dt}. \tag{1.4}$$

Oddly, torque has the same fundamental units as work and energy, $kg\ m^2\ s^{-2}$, but torque and work are very different (torque is a vector and work is a scalar, for example). Thus, torque is not measured in joules, but in newton meters (N m).

1.3.4 Unsteady Motion

When multiple forces are applied to an object, they can be added vectorially. If they sum to zero, there is no *net* force and so there is no acceleration. If they do not sum to zero, the object experiences a net force and it will accelerate. Constant or "uniform" acceleration is basic to particle or solid body mechanics and is generally treated as rather elementary. Cyclic or oscillating motion is more complex but still fairly straightforward for solid bodies.

In fluid mechanics, however, accelerations dramatically increase the complexity of analyses. The equations of motion for solid body kinematics are a small set of simple algebraic expressions of the form $v_x = v_{xo} + a_x t$.[c] In contrast, the equations of motion for a fluid are a large and intricate set of partial differential equations called the Navier–Stokes equations, and adding accelerations massively complicates any analysis. For constant or near-constant accelerations of solid bodies in fluids, simplified semiempirical approaches such as the added mass coefficient or the acceleration reaction (e.g., Daniel, 1983, 1984) can be used. Unsteady fluid motion, where both the fluid's velocity and acceleration change with time, can be especially troublesome. Consider the forces on the flapping wing of a hummingbird or on the beating tail of a swimming trout. Initial animal flight analyses used the "quasisteady" approximation, where researchers divided a wingbeat up into many small stages and treated each stage as a steady state. The whole wingbeat cycle was then modeled as a rapid succession of many steady states (Jensen, 1956; Norberg, 1976a). This approach produced reasonable first-approximation results, but later works showed that the quasisteady approximation is probably closest to reality for large, slow-flapping birds in forward flight and probably not very close at all for most hovering flight or for insects, small bats, and small birds (Norberg, 1976b; Ellington, 1984b; Dudley and Ellington, 1990; Alexander, 2002). Fortunately, researchers can use a dimensionless index, the Strouhal number (or a slight variation, the reduced frequency), to help decide whether the quasisteady approximation might be appropriate or not (see Section 3.5).

1.3.5 Derived Quantities Involving Force

Power is an important derived quantity that relates force, work, and energy. Power, P, is defined as work per unit time:

$$P = \frac{W}{t} \tag{1.5}$$

---

c. v_x is speed in direction x at time t, v_{xo} is initial speed, and a_x is acceleration in direction x.

And since work is (the magnitude of) force F times distance x, we can substitute for work and get

$$P = \frac{Fx}{t}. \qquad (1.6)$$

But x/t is speed, v, so power is also force times speed:

$$P = Fv. \qquad (1.7)$$

Thus, power is both the rate of doing work and the force needed to achieve a particular speed. Since energy must be expended to do work, power can also relate force to energy costs.

A force can be applied in any arbitrary direction on a solid body, but fluids can only apply force perpendicular to whatever solid surface they contact. Moreover, the force acts as if it were distributed continuously over this surface rather than concentrated at a point. This distributed force is pressure, formally the magnitude of the normal force per unit area of solid surface. Pressure is always perpendicular to the surface, so it does not have any direction independent from the surface and is thus a scalar. In the SI system, the pascal (P) is a derived unit of pressure, equal to $N\,m^{-2}$.

Most people are probably at least somewhat familiar with the concept of pressure. Stress (in the mechanics sense) is much less familiar. As we will see in Section 2.1, when forces are applied to solids, they set up distributed internal force fields called stresses. The direction of such a stress is determined by the way the load is applied, so again, stress has no direction independent of the loading conditions. Like pressure, stress takes the form of a force distribution continuously spread over an area, although in this case the area is a particular cross section of the object rather than surface area. Stress is thus measured as force per unit area and in the same unit as pressure, i.e., pascal in the SI system.

1.3.6 Mass and Weight

Few topics in physics lead to more confusion among nonphysicists than the distinction between mass and weight, especially when converting between systems of units. *Mass* is a measure of the amount of matter that makes up an object, whereas *weight* is a downward force due to gravity. My *mass* does not change even if gravity changes. If my mass is 65 kg on Earth, it will remain 65 kg if I am on the moon, on Mars, or even in outer space, regardless of the amount of gravity (or lack thereof). On the other hand, if my mass is 65 kg on Earth, then my weight will be 638 N (mass times the acceleration due to gravity, *mg*). My weight *does* vary with local gravity, so my weight will be 105 N on the moon, 241 N on Mars, and essentially 0 N on the International Space Station. Much of the confusion stems from the fact that we routinely weigh objects but typical weighing scales and balances automatically convert the weight into mass. When I place an object on the pan of an electronic scale (commonly but inaccurately called a "balance"), the scale measures how hard the object pushes down on the pan, which is a force (weight), yet the readout

FIGURE 1.2 The difference between weight and mass. Electronic top-loading scales ("balances") actually measure weight, even if they display in kilograms. On Earth, the upper electronic scale shows 1.0 kg for a particular object. The same scale shows 0.17 kg for the same object on the moon (lower left). A mechanical beam balance measures mass directly: a given object will balance against the same 1-kg reference mass on the Earth (upper right) and the on the moon (lower right). *Artist: Sara Taliaferro.*

may be in grams or kilograms—a mass (Fig. 1.2). The scale has a built-in calibration that assumes a particular constant value for the acceleration due to (Earth's) gravity, g, and automatically calculates mass, m, from weight, F_w, based on Newton's second law:

$$m = \frac{F_w}{g}. \tag{1.8}$$

Indeed, the value of g, by convention given as 9.81 m s^{-2}, varies slightly from place to place on the surface of the Earth, so the most sensitive electronic scales must be manually calibrated to account for very small differences in the local value of g every time they are moved. In contrast, one can measure mass directly. The humble pan balance—two pans suspended from each end of a pivoting beam—works by comparing the mass of an unknown object with the known mass of standard "weights." (Any kind of beam balance using sliding weights, including the humble laboratory triple beam balance, works on the same principle.) Yes, it requires gravity to work, and yes, it is based on comparing forces, but the exact strength of gravity *does not matter*: if an unknown mass is exactly balanced on a pan balance or beam balance by a standard 1 kg "weight" on Earth, the unknown object will still be exactly balanced by the same 1 kg "weight" on Mars or on the moon (Fig. 1.2). If the standard "weights" are calibrated in mass units, such a balance will give correct mass readings regardless of gravity, but it will give incorrect readings when g is different if the "weights" are calibrated in force (i.e., weight) units, such as pounds or newtons!

Biologists, in particular, have developed a bad habit of referring to mass units as weight, because for many biological purposes, they are interchangeable (e.g., Hill et al., 2016, pp. 178–180). In situations where calculations involve combinations of forces and masses, they are emphatically not interchangeable, and care must be taken to correctly convert weights to masses or vice versa. Fortunately, on Earth we are normally safe when we assume g is constant at 9.81 m s^{-2}, so mass and weight can easily be converted back and forth using Newton's second law. Nevertheless, referring to an object's "weight" in grams or kilograms is sloppy and unscientific; in the SI system, weight should be expressed in newtons.

Further contributing to the confusion is the fact that the SI system and the US customary system treat weight and mass differently. In the SI system, mass is a fundamental quantity based on the mass of the standard kilogram in Paris. Force is a derived quantity, the newton, equal to kg m s^{-2} from Newton's second law. (The "kilogram-force," kgf—the weight of an object with a mass of 1 kg—is sometimes used for convenience in commerce or informal settings. It is not a unit consistent with the standard SI system, and it must be converted into newtons [9.81 N per kgf] before applying any physics calculations based on first principles.) In the US customary system, the pound—specifically the common, everyday avoirdupois pound, lb—is a fundamental unit, and in traditional physics practice, it represents a weight (i.e., force) unit, making mass a derived unit. The mass unit in this system is the *slug*, equivalent to force divided by the acceleration of gravity ($\text{lb s}^2 \text{ ft}^{-1}$ in fundamental units). Unfortunately, the slug is even more unfamiliar to the average American than the newton is to the average European, so people often use the mass of an object with the weight of one pound as a mass unit, the pound-mass (lbm). Just as with the kgf, using pounds for both force and mass makes the system inconsistent, so lbm cannot be used in mechanics when lb is also used as a force. Ironically, the pound has been defined for over a century in grams: one pound is the weight of an object with a mass of just over 453 g.[d]

---

d. To be precise, 453.59237 g.

1.3.7 Other Physical Quantities

Many other physical quantities that play major roles in biology and other fields play little or no role in mechanics. Biologically important examples include temperature, light intensity, and voltage (electrical potential difference). An introductory physics textbook might devote as many or more of its pages to topics related to such quantities as to topics encompassed by mechanics. Topics such as entropy, electricity and magnetism, optics, and waves versus particles are all of fundamental importance in elementary physics, but they have little or no bearing on mechanics as we will use it in this book.

1.3.8 Units and the SI System

Mechanics, or indeed physics in general, is completely intolerant of inconsistent units. The modern SI system has been designed to be internally consistent so that conversion factors are not needed when correct units are used in equations derived from first principles. In the United States, the "US customary" units were used alongside "metric" units in science classes a couple of decades ago, but nowadays, all serious science is taught and performed in SI units. Students—particularly biology students—should be aware of units with a "metric system" aura that are not true SI units. The most widely used of these pseudometric units is the calorie (and its offspring, the kilocalorie = "large calorie" = food calorie), a unit for energy. Although the calorie is defined in terms of metric units, it is not a valid SI unit. Less common but still occasionally seen are mmHg or cm-H_2O for pressure. These units were developed for measuring convenience rather than consistency, and they cannot be used in standard mechanics equations. Simply becoming familiar with the common SI units described earlier in this chapter will prevent mistakes caused by inconsistent units.

FURTHER READING

General References

Alexander, R.M., 1968. Animal Mechanics. University of Washington Press, Seattle, 346 pp.

Vincent, J.F.V., 1982. Structural Biomaterials. Macmillan, London, 206 pp.

Vogel, S., 1981. Life in Moving Fluids: The Physical Biology of Flow, first ed. Willard Grant Press, Boston. 352 pp.

Wainwright, S.A., Biggs, W.D., Curry, J.D., Gosline, J.M., 1976. Mechanical Design in Organisms. John Wiley & Sons, New York, 423 pp.

Locomotion

Alexander, D.E., 2002. Nature's Flyers: Birds, Insects, and the Biomechanics of Flight. Johns Hopkins University Press, Baltimore, Maryland, 358 pp.

Bainbridge, R., 1958. The speed of swimming of fish as related to size and to the frequency and amplitude of the tail beat. Journal of Experimental Biology 35, 109–133.

Bennett, S.C., 2000. Pterosaur flight: the role of actinofibrils in wing function. Historical Biology 14, 255–284.

Brown, R.H.J., 1953. The flight of birds. 2. Wing function in relation to flight speed. Journal of Experimental Biology 30, 90–103.

Daniel, T.L., 1984. Unsteady aspects of aquatic locomotion. American Zoologist 24, 121–134.

Dudley, R., Ellington, C.P., 1990. Mechanics of forward flight in bumblebees. II. Quasi-steady lift and power requirements. Journal of Experimental Biology 148, 53–58.

Liu, Y.P., Sun, M., 2008. Wing kinematics measurement and aerodynamics of hovering droneflies. Journal of Experimental Biology 211, 2014–2025.

Material Properties

Alexander, R.M., 1962. Visco-elastic properties of body-wall of sea anemones. Journal of Experimental Biology 39, 373–386.

Denny, M.W., 1980. Silks: their properties and functions. Symposium of the Society for Experimental Biology 34, 247–272.

Ennos, A.R., van Casteren, A., 2010. Transverse stresses and modes of failure in tree branches and other beams. Proceedings of the Royal Society of London B: Biological Sciences 277, 1253–1258.

Sports Biomechanics

Cross, R.C., 2011. Physics of Baseball & Softball. Springer-Verlag, New York, 324 pp.

Higuchi, T., Nagami, T., Nakata, H., Watanabe, M., Isaka, T., Kanosue, K., 2016. Contribution of visual information about ball trajectory to baseball hitting accuracy. PLoS One 11, e0148498.

Chapter 2

Solid Materials

2.1 INTRODUCTION TO SOLIDS

The bodies of organisms usually contain quite a bit of water. Even the softest, squishiest organisms, however, have at least some solid components—skin, muscle—and many have rigid supporting material such as bone or wood. In this chapter, we will focus on the mechanics of solid materials and on the ways that many biological materials differ from typical building or "engineering" materials.

What exactly is a solid? Intuitively, you already know that if something is solid, it is not a fluid (liquid or gas). Solid materials maintain a fixed shape, whereas liquids and gases have no fixed shape and conform to the shape of whatever container they are in. Mechanically, solids are materials that resist being deformed, in contrast to fluids. Fluids do not resist being deformed, but instead resist the rate of being deformed. Engineers describe a solid's resistance to being deformed as its "elasticity," whereas they use "viscosity" to describe a fluid's resistance to its rate of being deformed. (The technical meaning of "viscosity" is thus fairly close to its everyday meaning, but we will see that the technical meaning of "elasticity" is somewhat different from its everyday meaning.) A great deal of the study of solid mechanics revolves around exactly how a material deforms—or breaks—in response to various kinds of loads.

The materials used by engineers tend to fit into the "solid" and "fluid" categories neatly and unequivocally; at room temperature, steel is a solid and gasoline is a fluid. Most biological solid materials, however, have at least a little fluidlike behavior and are thus technically viscoelastic; because viscoelastic properties combine solid and fluid aspects, we need to understand both conventional solids and conventional fluids before tackling viscoelastic solids in Chapter 4. Nevertheless, the hardest biological solids do act much like engineering solids, and the sophisticated methods that engineers have developed to analyze solid materials can be usefully applied to many stiff biological materials.

Nature's Machines. http://dx.doi.org/10.1016/B978-0-12-804404-9.00002-5

2.2 LOADING, DEFORMATION, STRESS, AND STRAIN

2.2.1 Loads and Deformations

According to Newton's third law, if I push on a brick wall with a force of 100 N, the brick wall pushes back with the exact same force. My muscles consume energy to generate that force, but where does the 100-N force of the brick wall come from? In a word, it is deformation. By pushing on the wall, I deform it ever so slightly, squeezing the molecules of the bricks a tiny bit closer together and that distortion of the molecular arrangement produces the reactive push. The more the molecules are distorted, the greater the reactive force produced. Some materials, such as brick, push back with so little distortion, the actual deformation is tiny and hard to measure. Other deformations are easily visible, such as the deformation of a bow when the archer pulls back on the bowstring. The deformation may be very small, but whenever a force ("load") is applied to a solid, some deformation must occur.

Solids can be deformed in different ways. If I use a rope to hang a heavy bucket of rocks from a tree branch, the rope is loaded in *tension* (Fig. 2.1A). The rope elongates slightly and this stretching produces the equal and opposite force to resist the weight or downward force of the bucket. If I place a block of wood on a concrete floor and then stand on the wood block, I am loading the block in *compression* (Fig. 2.1B), where the top and bottom of the block are forced a tiny bit closer together. This shortening again produces the equal and opposite reaction of the wood to support my weight. If I place one lower edge of the block of wood against a low obstacle—a ruler glued flat to the floor—and push on the upper, opposite edge, the block deforms in *shear* (Fig. 2.1C). Shear is produced by parallel but offset forces that tend to slide parallel faces of the object in opposite directions. All solid deformations consist of the material's response to some combination of these three types of loading.

Sometimes the deformation can be quite apparent. Stretching a rubber band is an extreme example. Rope may not appear to be particularly stretchy, but whether made of natural or synthetic material, it will elongate at least some in tension. If a swing seat is suspended from a high tree branch by a pair of ropes,

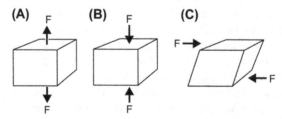

FIGURE 2.1 Three ways in which forces can be applied to deform a solid. (A) Tension; (B) compression; (C) shear. *F*, force. *From Alexander, D.E., 2016. The biomechanics of solids and fluids: the physics of life. European Journal of Physics 37, 053001. Artist: Sara Taliaferro,* © *European Physical Society. Reproduced by permission of IOP Publishing. All rights reserved.*

the swing will sag noticeably when a person sits on it. Even if the rope only elongates $\frac{1}{2}\%$, that elongation would produce a 5 cm sag in a 10-m rope. On the other hand, depending on the load and the stiffness of the material, the deformation may be minute. When I stand on a concrete floor, the microscopic deformation from my weight would be challenging to measure. Of course, with a large enough load, even the deformation of concrete can become apparent, such as when heavy trucks drive over concrete bridges and viaducts.

Some solids can resist great loading in tension but offer negligible resistance in compression. When formed into cords or sheets, such materials make up tension-resisting engineered structures such as rope or biological counterparts such as tendon. These can be termed *tensile* materials. In contrast, engineers make abundant use of materials that resist compression mightily but are rather weak in tension, such as masonry and unreinforced concrete. These materials are "hard" in the everyday sense, but have a dismaying tendency to fall apart under modest tension; living organisms do not seem to make much use of such materials. Yet other materials do more or less equally well resisting both tension and compression. In the technological realm, these include metals such as steel and aluminum alloys with counterparts such as bone, wood, and arthropod cuticle made by organisms.

A variety of biological materials have properties that fall in between the tensile materials that make up ropes and tendons and the rigid materials such as steel and bone. They have some flexibility, but also some ability to resist tension and compression. Wainwright et al. (1982) called this intermediate category *pliant* materials, and it includes floppy materials such as jellyfish mesoglea ("jelly"), stretchy materials such as resilin and other rubbery proteins, and stiffer but still flexible materials such as cartilage. These pliant materials all show some degree of time-dependent or viscoelastic behavior (see Chapter 4) in contrast to rigid biomaterials[a] such as wood and bone, which do not.

2.2.2 Stress and Strain

One of the most important properties of any solid material is how it behaves when loaded, i.e., when a force is applied. We could, for example, take a sample piece of some solid, attach one end to a fixed support, and pull on the other end (Fig. 2.2A). We can then measure how much the sample elongates as we apply greater and greater force until the sample breaks or "fails." If we then make a graph of force applied, F, versus sample extension, ΔL, we might get a graph looking like that of Fig. 2.2B. Such a graph shows the amount of force required to stretch that particular sample a given amount, but it cannot tell us anything about larger, or smaller, or differently shaped samples. If we want to know about the properties of the material independent of shape and size, we

---

a. I am using "biomaterial" to mean material naturally produced by organisms. In biomedical engineering, the term is often used to mean synthetic material for making implantable devices.

FIGURE 2.2   Terminology for a typical solid sample being tested in tension. (A) A force F is applied in tension to a material sample with an initial length L_0, which extends the sample length by ΔL. (B) A plot of force F versus extension ΔL. The force is increased until the sample breaks, or fails, at x (by convention, the curve normally just ends at the point of failure). (C) The same force-extension data, but normalized as stress and strain. At extensions or strains above the yield stress, irreversible or plastic deformation occurs. *Redrawn by Sara Taliaferro from Alexander, D.E., 2016. The biomechanics of solids and fluids: the physics of life. European Journal of Physics 37, 053001, © European Physical Society. Reproduced by permission of IOP Publishing. All rights reserved.*

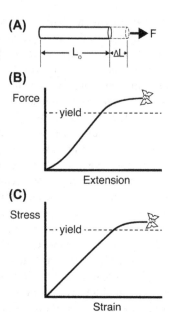

need to normalize the force and extension. We normalize force by dividing force by the cross-sectional area, S, of the sample, which gives the *stress*, σ:

$$\sigma = \frac{F}{S}. \tag{2.1}$$

Stress is force per unit area, so it can be measured as $N\ m^{-2}$ or Pa. The pascal is inconveniently small for practical measurements of real materials, so megapascals (MPa) or gigapascals (GPa) are commonly used. (Although not a standard SI unit, the newton per square millimeter, $N\ mm^{-2}$, is sometimes used in material testing research, and conveniently it turns out to be identical with the MPa.)

Rather than plotting the total length of the sample across a range of stresses, we use the proportional change in length relative to the original unstressed length or *strain*, ε. Engineers mostly work with materials that do not stretch very much before they fail, so they use the ratio of the change in length, ΔL, to the original length, L_0. Engineering, nominal, or Cauchy strain[b] is thus given by

$$\varepsilon_{en} = \frac{\Delta L}{L_0}. \tag{2.2a}$$

---

b. Engineering strain is sometimes expressed as a percentage, i.e., the proportion as shown in Eq. (2.2a) times 100%. Generally on graphs, engineering strains can be assumed to be the proportion as in Eq. (2.2a) unless a percent sign is explicitly shown.

When strains exceed 10% of the original length, however, engineering strain no longer accurately represents the incremental length change with increasing stress, so for very extensible materials, scientists use the true or natural strain, ε_t (Beer and Johnston, 1981, p. 38):

$$\varepsilon_t = \int_{L_0}^{L} \frac{dL}{L} = \ln\left(\frac{L}{L_0}\right) = \ln\left(\frac{\Delta L + L_0}{L_0}\right). \tag{2.2b}$$

At extensions of less than 10%, engineering and true strains are almost identical, but they begin to diverge substantially above 10%. Both versions of strain are ratios, and, thus dimensionless.

2.2.3 Information From Stress–Strain Curves

When data from a force-extension test are converted into stress and strain, they can be plotted on a stress–strain graph (Fig. 2.2C). Such a stress–strain curve gives a number of properties of the material independent of the geometry of the sample. (The convention of putting stress on the y-axis and strain on the x-axis may seem perplexing, see Box 2.1.)

BOX 2.1 Axes of the Stress–Strain Curve

The way I have described the measuring process to produce stress and strain data, one might argue that the axes should be reversed: if we apply a stress and measure the resulting strain, shouldn't stress be the independent variable and hence on the x-axis, with strain as dependent variable on the y-axis? A materials scientist would argue that from the material sample's perspective, a deformation is being applied to the material, which generates an internal force distribution, i.e., stress, in the sample. So from the sample's point of view, elongation is the independent variable on which the stress depends. In fact, some material testing devices actually work this way, by applying a known extension or compression and measuring the resulting tensile or compressive force. Because the method I originally described—applying a known force and measuring the deformation—is simpler both heuristically and in practice, it is the usual method used to introduce stress and strain. Thus, students may be left with the mistaken impression that engineers do not follow the conventions for independent and dependent variables on graphs.

One of the most fundamental mechanical properties of a solid is the slope of the linear part of the stress–strain curve. This slope is called the "Young's modulus" or "modulus of elasticity," E (Beer and Johnston, 1981, p. 38). Note that the Young's modulus is given by $E = \sigma/\varepsilon$, and since strain (ε) is dimensionless, Young's modulus has the same unit as stress, Pa (or MPa or GPa). Materials with a steep slope—high Young's modulus—do not extend much even under heavy loads, and these are called stiff materials. Materials with a flatter slope—lower Young's modulus—stretch more for a given load and are less stiff or more *compliant* (compliance being the reciprocal of stiffness).[c]

The upper right-hand end of the stress–strain curve indicates two other properties. The height of the end of the curve gives the material's *strength* (sometimes called "ultimate strength") (Ennos, 2012, p. 9). The strength is thus the value of stress at failure, the maximum force per unit area the material can sustain before breaking. Since a material's strength is a particular value of stress, strength is yet another property measured in the same unit as stress, i.e., pascal. The horizontal distance of the end of the stress–strain curve from the origin gives the material's *extensibility* (Ennos, 2012, p. 9). Some brittle materials, glass for example, have very low extensibility: they fail with very little extension, a tiny fraction of a percent. In contrast, many biological materials—human skin, spider silk, expansion joints in insect exoskeletons—extend anywhere from 50 to several hundred percent beyond their original unstressed length.

The area under the stress–strain curve displays yet another property. This area represents the amount of energy stored by the material as it deforms. This energy goes by various names, including *strain energy* or *work of extension* (or *compression*) (Vogel, 2013, p. 303). If we extend this area out to the point of failure, it becomes the *toughness* (Fig. 2.3A).[d] This quantity can be viewed as either the amount of work done to deform the object or the amount of energy absorbed by the material's chemical bonds as it deforms. Just as we use stress as a normalized form of the force, the work of extension or toughness is actually a scale-independent form of the strain energy, energy per unit volume. It could, for example, be measured as the total work done on a sample divided by the volume of the sample. (Strictly speaking, "strain energy" or "work of extension" should include the phrase "per unit volume," but it almost never does in the biomechanics literature; Wainwright et al., 1976; Vogel, 2013). I leave it to readers to demonstrate themselves that energy per unit volume comes out in units of $N\,m^{-2}$, yet another material property measured in pascals!

Toughness and strength need not be correlated. Brittle materials by definition have low extensibility, so even when a brittle material such as cast iron

---

c. Note that the everyday meaning of "elasticity" as stretchiness is rather different from its use in materials science, where the "elastic (Young's) modulus" refers to stiffness, and "elasticity" refers to an object's ability to return to its original shape after being deformed. Steel ball bearings are elastic in the technical sense but not in the everyday sense.

d. In material science, "toughness" is used as the opposite of "brittleness." It is usually defined as above (work per unit volume) and sometimes called "modulus of toughness" (Beer and Johnston, 1981). Engineers also use various specific tests for toughness (e.g., Charpy and Izod), each with its own specialized units. Confusingly, "toughness" is sometimes used synonymously with "work of fracture," the energy per unit surface area needed to form new surface during fracture, see Section 2.2.1 (Vogel, 2013).

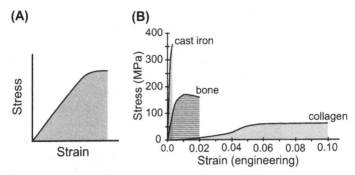

FIGURE 2.3 Strain energy storage or work of extension. (A) The area under the stress–strain curve (shaded) represents strain energy stored per unit volume as the sample deforms; if stressed to failure, this area is the toughness. (B) Brittle materials such as cast iron have low toughness compared to many biological structural materials; pliant biomaterials such as collagen can be tougher than rigid supporting biomaterials such as bone.

has high ultimate strength, it has very little area under its stress–strain curve (Fig. 2.3B). Cortical bone, which has about half the ultimate strength of cast iron, can store several times more energy before failing. And the extensibility of collagen, with less than half the strength of bone, gives collagen much greater toughness even than bone, let alone cast iron (Fig. 2.3B).

Typical engineered materials are either entirely (e.g., glass, ceramics) or partly (e.g., steel and aluminum alloys) linearly elastic: some, if not all, of the stress–strain curve is a straight line (Fig. 2.4). When such materials are loaded so that they remain on the linear part of the curve, they return completely to

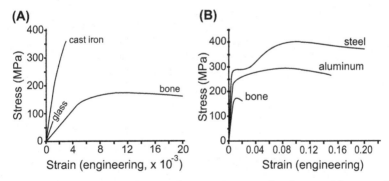

FIGURE 2.4 Sample stress–strain curves for some common engineering materials compared to a biological material (cow cortical bone). (A) Glass and cast iron show essentially no plastic behavior, whereas bone extends significantly after its yield point. (B) Low-carbon steel, or "mild" steel, and structural aluminum alloy have both great strength and considerable plastic deformation in comparison to bone. (Note the different horizontal axis scales in (A) and (B).) These and all other stress–strain curves for particular materials are composites based on many sources and are somewhat idealized; they do not represent specific data and are for comparison purposes only.

their original, undeformed shape. In the process, they return all of the energy that they absorbed while being deformed.[e] Such linearly elastic behavior is also called Hookean behavior, from Hooke's law ("deflection is directly proportional to force applied") (Gordon, 1976, p. 36). Most engineered materials are completely Hookean or have substantial Hookean ranges, and biological materials with a substantial range of Hookean elasticity, in addition to bone, include keratin from feathers and hooves, and some kinds of cartilage. Otherwise, most biological materials do not display Hookean behavior; they do not have an obvious linear region on their stress—strain curve. For those biological materials lacking a linearly elastic range, the standard version of Young's modulus cannot be used, so as we will see in the next section, a modified version of the modulus of elasticity must be defined.

The stress—strain curves in Figs. 2.2 and 2.4B display a *yield stress*, beyond which the curve is no longer linear. When a material is stressed beyond the yield stress, some of the deformation becomes permanent. This permanent deformation is called *plastic* deformation, and any work that goes into plastic deformation will not be returned when the load is removed. In fact, the vast majority of biological materials show complex, non-Hookean stress—strain behavior, and they vary enormously in how much energy they return when unloaded, as we will see in Section 2.2.6. Vertebrate compact bone is unusual for a biomaterial in having both a significant linear stress—strain region as well as a substantial region of yield beyond the linear zone (Fig. 2.4). Bone has considerable postyield strain compared to brittle materials such as glass (Fig. 2.4A), but bone's plastic behavior is relatively modest in comparison to some metals, such as mild steel or aluminum alloy (Fig. 2.4B).

Whereas Young's modulus applies to both tension and compression, a different relationship is needed for shear. The shear stress,[f] τ, is the force applied, F_s, divided by the area *parallel to* the force, S_s ($\tau = F_s/S_s$), and shear strain is the ratio of the amount of deflection parallel to the force, ΔL_x, to the object's unstressed length perpendicular to the force L_{y0}, i.e., tan γ (Fig. 2.5). (At low shear strains, the deflection angle γ itself, in radians, will be essentially the same as the tan γ, so the angle itself is usually used as shear strain for

---

e. In reality, a bit of the work of straining the material is lost as heat, so either the material will suffer imperceptible permanent deformation, or enough heat is absorbed from the environment to return the material to its undeformed shape. Either way, such a material, for all practical purposes, appears to return essentially all of its strain energy when unloaded.

f. The shear stress and the tensile-compressive stress are actually components of a *stress tensor*, σ_{ij}, which fully describes the stress in a solid in three dimensions. The stress tensor has nine components, a tensile-compressive stress and an orthogonal pair of shear stresses for each of the three dimensions. For our purposes, looking at one tensile-compressive or shear stress at a time is adequate; a full mathematical description of the mechanical behavior of a solid using the stress tensor is beyond the scope of this book.

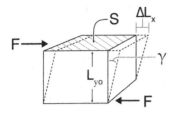

FIGURE 2.5 Terminology for shear. Equal and opposite forces (F) are applied, causing angular distortion γ. Strain is given by the ratio of distortion ΔL_x to unstressed thickness L_{y0}. Shear stress is normalized by the area S parallel to the forces. *From Alexander, D.E., 2016. The biomechanics of solids and fluids: the physics of life. European Journal of Physics 37, 053001. Artist: Sara Taliaferro,* © *European Physical Society. Reproduced by permission of IOP Publishing. All rights reserved.*

stiff materials.) The shear modulus, G, takes the same form as Young's modulus:

$$G = \frac{\tau}{\gamma}. \tag{2.3}$$

The dimensions are the same as for tension and compression: τ and G are force per unit area (so again Pa in the SI system), and both tan γ and γ (measured in radians) are again ratios and thus dimensionless.

Material samples can be tested in shear just as in tension or compression. The shear–stress/shear–strain curves will show the shear modulus, as well as the shear versions of properties we saw on tensile stress–strain curves such as strength and toughness.

2.2.4 Nonlinearity

Many common building and structural materials possess a clearly defined linearly elastic range followed by a clearly defined plastic range, as in Fig. 2.4B. Above the yield stress, the material deforms permanently: if the load is removed before failure, the material will only return partway to its original shape. The part of the stress–strain curve beyond the yield point is the plastic region of the curve. Some materials such as cast iron or glass have little or no plastic behavior, they deform elastically right up to failure (Fig. 2.4A). Others, such as mild structural steel or aluminum alloy may actually deform more plastically than elastically (Fig. 2.4B). Note how a significant plastic region can dramatically increase the area under the stress–strain curve, producing a severalfold increase in energy storage or toughness. This plastic region can act as a shock absorber or safety relief; by absorbing a lot of energy, the plastic deformation can prevent failure even at unusually high loads. Some materials, such as wet clay, show no elastic deformation at all. Their deformation is entirely plastic, so all deformation is permanent. (Note that, even more than "elastic," the technical meaning of "plastic" in mechanics is entirely different from its meaning in everyday usage.)

Mammalian bone is rather rare among biological materials in having a substantial linearly elastic initial strain. Most biological materials have mostly or completely nonlinear stress–strain curves. Consider collagen, a class of common structural proteins throughout the animal kingdom. Collagen molecules have a very ropelike form, consisting of three helical polypeptide chains twisted into a triple helix; these molecules are usually assembled into tubular fibrils that make up the basic structural unit of most collagenous structures (Wainwright et al., 1982, pp. 83–87). Collagen is used in tensile structures such as tendons, compression-resistant materials such as cartilage, and pliant gels such as sea anemone mesoglea. Collagen has a distinctly J-shaped stress–strain curve as shown in Fig. 2.6. Note that at the low end, little force is needed to stretch the material, but above a certain strain level the curve rises steeply, indicating ever-growing stress needed to stretch it. The standard interpretation of this behavior is that the collagen fibers are kinked or crimped when relaxed, so the first part of the curve represents straightening kinked fibers, whereas the later, steeper part of the curve is where chemical bonds between and within the fibers start to stretch (Wainwright et al., 1976, p. 89; Vincent, 1982, p. 60). Many biological materials have a J-shaped stress–strain curve, including some kinds of spider silk, flexible cuticle in insect joints, and both skin and arterial walls in mammals.

Collagen can be strained up to about 5% or 10% and still return a very high proportion of the energy absorbed, over 90%. The proportion of energy returned is called the *resilience*. In addition to collagen, several biological materials have evolved to work as springs with very high resilience. These "protein rubbers" work over far larger strains than collagen, up to $\varepsilon_t = 1.0$ (100% extension or more; see Box 2.2). Their resilience ranges from about 76% for elastin (from mammals) (Wainwright et al., 1976, p. 118) to over 95% for resilin (from insects) (Weis-Fogh, 1960). These protein rubbers tend to have a stress–strain curve with a low, almost linear initial region over a very

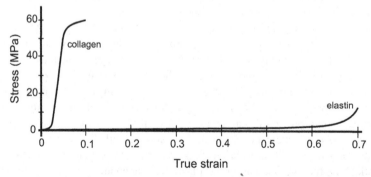

FIGURE 2.6   Stress–strain curve for two non-Hookean biological materials. (Note different scale of strain axis compared to Figs. 2.3 and 2.4.) *From Alexander, D.E., 2016. The biomechanics of solids and fluids: the physics of life. European Journal of Physics 37, 053001,* © *European Physical Society. Reproduced by permission of IOP Publishing. All rights reserved.*

BOX 2.2 Rubber Elasticity

Scientists figured out the mechanics of rubber elasticity based on natural and artificial rubbers. The so-called "natural rubber" is made from the sap (latex) of the rubber tree; it is not at all rubbery when used by the tree. It becomes rubbery in the vulcanization process, where widely separated bonds ("cross-links") form between the long, floppy, spaghetti-like molecules of the latex. These bonds constrain the molecules enough to change the material into a solid, but they allow lots of thermal movement of the molecules between the bonds. When vulcanized rubber is stretched, the scattered, randomly arranged bonds are pulled into a more orderly arrangement, aligned more parallel to the stress, and also closer together perpendicular to the stress. The thermal movements of the molecules resist this loss of randomness (loss of entropy) weakly at low strains and allow great elongation; at high strains, the stiffness rises sharply as the molecules and their bonds themselves start to stretch. This molecular behavior produces the pronounced J-shaped stress–strain curve of rubbery materials (Ennos, 2012, pp. 32–35). Because the rubber molecules thrash about more frantically at higher temperatures, making them more resistant to being stretched, rubbers are unusual in that they get stiffer rather than softer at high temperatures. Two of the three protein rubbers found in nature, resilin from insects and abductin from bivalve mollusks, seem to get their rubbery properties from this classical amorphous rubber elasticity mechanism (Vincent, 1990, p. 66).

Elastin is the rubbery protein found in vertebrates. Elastin makes your large arteries stretchy and allows them to smooth out the pulses in blood flow produced by the beating heart (see Section 4.1). Elastin does not fit the classic amorphous rubber molecular arrangement. Instead, elastin molecules form helical fibers. Current evidence suggests that these fibers are triple helices of helical proteins, and these helical proteins have regularly spaced kinks or bends ("β-turns") that allow free rotation. Straightening these bends or pulling them more into alignment gives the fibers their rubbery elasticity (Ennos, 2012, pp. 38–41).

Unlike natural or synthetic rubbers, all protein rubbers must be well hydrated to show rubbery properties. When dry, they become stiff, crystalline solids. They absorb water and swell, and this water facilitates the thermodynamic movements of molecules. The water is necessary for the proteins to show the high strains and high resilience needed for the material to function properly (Wainwright et al., 1982, p. 111).

large strain range, followed by a relatively rapid, dramatic increase in slope (Fig. 2.6). They thus have very low stiffness over a large extension, and then with just a bit more extension, they become very stiff. This combination of high extensibility and high resilience makes protein rubbers well suited to acting as energy storage springs. Some insects have resilin in their wing hinges to store and return energy at the top or bottom of a wing stroke, and cattle have elastin in a neck ligament that helps support the weight of their heads.

Most pliant or tensile biomaterials are not as resilient as collagen or resilin; when they stretch and relax, they may not show much evidence of permanent deformation, yet they return a significantly lower amount of energy than they absorb. When this happens, the material follows a different, lower curve on the stress—strain graph than it followed while being stretched (Fig. 2.7). Following different curves when being strained and relaxing is called *hysteresis* (Vincent, 1990, pp. 32, 49). The area under the relaxing (unloading or returning) curve is the strain energy returned, so the area between the loading and unloading curve represents the amount of strain energy not returned. This energy is partly absorbed by reversible molecular reconfigurations (Vincent, 1982, p. 49) and partly lost as heat. Such a loading/unloading curve also gives the resilience of the material: the area under the loading curve gives the total energy absorbed, so the percentage of that total accounted for by the area under the unloading curve is the resilience. Some biomaterials seem to have evolved high hysteresis for useful function, such as some kinds of spider silk. This property allows them to capture insects by stretching on impact and absorbing the impact energy, rather than rebounding and flinging the insect back out of the web (Denny, 1980). A spider web with the properties of a trampoline would not catch many prey! (For details, see Section 4.1.4.)

2.2.5 Strength Versus Toughness

Strength need not be at all related to toughness (or strain energy storage). This may sound counterintuitive, but the apparent riddle may be largely due to the way we use the terms in everyday speech. Strength in the mechanical sense simply means the maximum stress a material withstands before failure. High-strength materials tend to have high Young's moduli (steep stress—strain curves) and many show little yield or plastic behavior. Materials such as glass and cast iron show this pattern (Fig. 2.8). Cast iron has a very high strength by

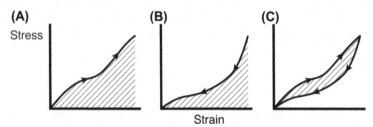

FIGURE 2.7 Loading and unloading for a hypothetical biological material. (A) Loading stress—strain curve; shaded area equals strain energy stored. (B) Unloading stress—strain curve; shaded area equals strain energy returned as stress is removed. (C) Hysteresis between loading and unloading curves; shaded area shows energy lost as heat. *From Alexander, D.E., 2016. The biomechanics of solids and fluids: the physics of life. European Journal of Physics 37, 053001. Artist: Sara Taliaferro, © European Physical Society. Reproduced by permission of IOP Publishing. All rights reserved.*

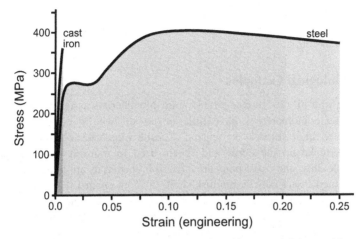

FIGURE 2.8   Difference in strain energy storage (toughness) between a brittle material, cast iron, and a tough material, mild steel. Shaded area under curves shows strain energy storage.

biological standards, but because its stress—strain curve is steep and has no plastic region, the area under its stress—strain curve is relatively small and has low toughness. Such materials are *brittle*: they do not absorb much energy before failure and so are particularly susceptible to fracture from shocks or sudden load increases. "Brittle" here has the same meaning as in everyday usage. A dry twig is brittle—it fails suddenly with modest bending—whereas a green twig is tough, failing gradually and requiring much greater bending to get even partial failure.

Some so-called ductile materials,[g] particularly mild steels and aluminum alloys, have a combination of great strength and high toughness (Fig. 2.8). These properties are why we make such great use of them in our technology. In biology, however, tough materials tend to have low strength (e.g., collagen, spider silk), whereas strong materials tend to be more brittle (e.g., tooth enamel, mollusk shells). Mammalian bone is an exception, being strong for a biomaterial with fairly high toughness (Figs. 2.3 and 2.4). Even bone's toughness is less than one-fourth of the toughness of the protein rubbers (e.g., abductin) and barely 15% of that of spider frame silk (Vogel, 2013, p. 304). Moreover, Vogel suggested that a more appropriate comparison for biological purposes is strain energy to failure *per unit mass* rather than per unit volume (Vogel, 2013, p. 304). On that basis, collagen and a stemlike structure of a giant seaweed (the stipe of bull kelp, a brown macroalgae) are an order of magnitude tougher than steel and slightly tougher even than bone. The great extensibility of such biomaterials means that they can absorb a lot of energy

---

g. "Ductile" is used in materials science to refer to metals with considerable extension after the yield point.

before failing, so even soft, flexible structures such as tendons, mammalian skin, and spider webs can be quite difficult to tear or break, particularly on a per-unit-mass basis.

2.2.6 Biological Examples

Wainwright et al. (1976) categorized solid biomaterials as tensile, pliant, or rigid. Tensile biomaterials have little or no stiffness in compression but moderate to high stiffness in tension. Tensile biomaterials—most notably, collagen—make up the cords and sheets used to transmit pulls, tie parts together flexibly, and resist punctures. Rigid biomaterials are those like bone, wood, tanned insect cuticle, mollusk shell, or tooth enamel that are hard and resist compression, and generally do not deform much in normal operation. Wainwright et al. described pliant biomaterials as supportive materials that deform substantially in normal operation. Some are largely a single component, such as the resilin pads found in the wing hinges and jumping systems of many insects. Other pliant materials are *composites*—made up of two or more components—with stiff tensile fibers such as collagen or chitin embedded in a gel, usually of protein or mucopolysaccharide suspended in water. The skin and arterial walls of mammals are some of the best-studied pliant composites.

Biomaterials are usually *anisotropic*—they have different properties in different directions—in contrast to most engineered building materials, which are *isotropic*, i.e., the same in any direction. A sample of steel has the same Young's modulus and toughness in any direction; this contrasts distinctly with wood, which can be very tough parallel to the grain, but barely 10% as tough across the grain (Vogel, 2013, p. 307). This difference becomes quite obvious when splitting firewood. A wedge applied to the end of a log can be effective because it separates the wood across the grain (between fibers). A wedge applied to the side of a log, perpendicular to the grain, is ineffective—not only will it fail to split the wood, but it will also rebound dangerously after a sharp sledgehammer blow, threatening the would-be woodsman. Indeed, wood parallel to the grain can approach steel in toughness (on the conventional per-unit-volume basis) and can significantly exceed steel on a per-unit-mass basis. A word of caution: the properties of living or "green" wood are significantly different from those of kiln-dried wood (lumber) prepared for structural applications. A great deal of information on the mechanical properties of dried lumber is available due to its commercial value, but relatively little is available on wood in the live state. Based on the behavior of dried and green twigs, I feel safe in asserting that trunks of living trees are substantially tougher than commercial lumber made from those trees.

Many animals make use of the energy storage ability of resilient biomaterials. Kangaroos use their ankle tendons (mostly collagen) as springs to store energy on landing and return it for the next jump (Alexander and Vernon, 1975). Biewener et al. (1981) calculated that this saves the kangaroo up to 35%

of the energy it needs for locomotion. Migratory locusts (Weis-Fogh, 1960) have pads of resilin in their wing hinges that do essentially the same thing during wing flapping as kangaroo tendons do in hopping, again leading to energy savings and possibly some automatic movement control. Fleas cannot contract their muscles fast enough to jump because of their small size. Instead, they use their muscles to bend their legs and deform a pad of resilin (they can do this relatively slowly), and then the flea releases a catch and uses the stored energy of the resilin pad to propel its jump at the phenomenal acceleration of 1330 m s^{-2}, over 100 times the acceleration of gravity (Bennet-Clark and Lucey, 1967)! These energy storage mechanisms will be described in more detail in the context of locomotion and jump height in Chapter 5.

Resilin nicely illustrates how nature evolves solutions that take a different approach from engineering technology. If humans need a spring, we often make one from a coil of stiff steel wire. A straight wire would have very little "give" under load, but by coiling it we can get a lot of extensibility—by slightly straightening the coils—using a rather inextensible material. In contrast, animals have evolved several rubber-like proteins, including resilin, that often function as springs. Where a human designer might take a standard material and give it a new form to achieve new properties—forming a steel wire into a coil to make it stretchy—sometimes nature simply evolves a novel material with the beneficial properties. Thus, different rubber-like proteins used as springs have evolved independently at least three times: resilin in insects, elastin in mammals, and abductin in mollusks. Of course, evolution can also take the same approach as human engineers, modifying the shape to get different properties. Spongy bone in mammals, for example, is not as strong as compact bone, but spongy bone is springier, lighter, and possibly tougher than compact bone, even though the bone tissue is essentially the same at the microscopic level. (Spongy bone, as the name suggests, is an open, three-dimensional (3D) meshwork in contrast to compact or cortical bone, which looks solid to the naked eye.)

Organisms have not evolved the ability to use pure metals (or alloys) for structural support, but they have evolved a wide variety of rigid supporting materials. Many of these—bone, mollusk shell, coral skeleton—use hard mineral crystals, such as calcium carbonate, embedded in a protein matrix to form biological ceramics. Others use stiff, inextensible fibers in a complex composite material, such as cellulose in wood or chitin in insect cuticle. The fiber arrangement determines much about the mechanical properties of such composites. Wood fibers are generally in parallel arrays, so wood's properties differ greatly parallel and perpendicular to the grain. Insect cuticle, in contrast, has the fibers in many layers, where the fibers in each layer are parallel but in a different direction from the fibers in layers above and below, much like plywood (Fig. 2.9). This arrangement gives it great toughness (and strength) in any direction in the plane of the layers. Many aquatic crustaceans start with the basic layered chitin composite of insect cuticle and add mineral crystals such as calcium carbonate to form very hard, rigid, exoskeletal plates that still retain

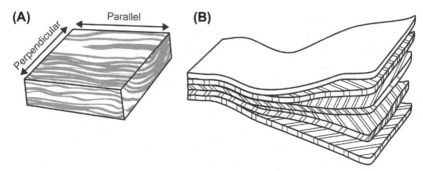

FIGURE 2.9 Fiber directions in wood and insect cuticle. (A) Wood, with fibers parallel to grain. (B) Insect cuticle forms in layers, with each layer having a different fiber orientation. *Artist: Sara Taliaferro.*

the toughness of chitin. Such mineral crystals add substantially to the exoskeleton's weight, but this is partly offset under water by buoyancy due to displacing water. For insects, particularly the majority of which can fly, a mineralized cuticle produces too much of a weight penalty, so insects make do with chemically hardening ("tanning") their protein–chitin composite exoskeleton.

Table 2.1 shows some material properties for a variety of biological and engineered materials. The Young's modulus values for the biological materials are tangent moduli and should be taken with a grain of salt given natural variation among individuals and differences in experimental protocols. Note that elastin and rubber both have low E but very high extensibility, so they are fairly tough. The structure of their polymer molecules allows lots of stretching and reorienting before failure. Materials with low extensibility are brittle; they can be strong and brittle, such as tooth enamel or glass (as in Table 2.1), or weak and brittle, such as gelatin or saltine crackers. As structural materials go, cartilage is fairly weak and brittle. Tendon (mostly collagen) is not very strong, but its extensibility makes it tough. It is so extensible because its collagen fibers are crimped when relaxed, and the initial low slope of the stress–strain curve is due to the straightening of these fibers. Wood (parallel to the grain) and bone are both fairly strong and tough. The very high strength and toughness of steel shows why metals are a favored structural material for engineered objects (and the value in Table 2.1 is just inexpensive "mild" or low-carbon steel used for everyday structural purposes).

2.3 FAILURE AND HOW TO PREVENT IT

When designing structures, engineers put a great deal of thought and analysis to avoid breakage or mechanical failure. A broken device is not only nonfunctional, a broken bridge or airplane can result in loss of life. So too in

TABLE 2.1 Material Properties of a Selection of Engineered and Biological Solids

	E (MPa)	Strength (MPa)	Extensibility	Toughness (MJ m^{-3})
Elastin	1.0	1.3	1.2	
Rubber	3.0	7	3	10
Cartilage	20	(1)	0.08	
Spider silk (capture thread)	1.0	534	1.72	75
Brick	700	5 (36)		
Tendon	1,500	100	0.10	2.8
Insect cuticle	5,000	100	0.02	0.03
Tree trunk (parallel to grain)	6,400	200	0.025[a]	5.0[a]
Spider silk (web frame, dragline)	8,000	900–1,500	0.2–0.33	100–190
Silk (silkworm, textile)	8,900 –17,000	120–230	0.25	
Cortical (dense or compact) bone	18,000	200 (170)	0.02	3.0
Tooth enamel	60,000	35 (200)	0.0005	0.02
Glass	60,900	70	0.001	0.03
Mild steel	200,000	400–900	0.2–0.3	90–120

Values of E for biological materials are tangent moduli. Strength values represent tensile strength (except values in parentheses, which represent compressive strength) and should be considered as estimates for comparison only. Toughness is equivalent to strain energy storage or maximum work of extension.
[a]Estimated, from steamed timber.

Data from Wainwright, S.A., Biggs, W.D., Curry, J.D., Gosline, J.M., 1982. Mechanical Design in Organisms (paperback ed.). Princeton University Press, Princeton, New Jersey, 423 pp., Alexander, R.M., 1983. Animal Mechanics, second ed. Blackwell Scientific Publications, Oxford, 301 pp., Currey, J.D., 1984. The Mechanical Adaptations of Bones. Princeton University Press, Princeton, NJ, 294 pp., Vincent, J.F.V., 1990. Structural Biomaterials. Princeton University Press, Princeton, New Jersey, 206 pp., Pérez-Rigueiro, J., Viney, C., Llorca, J., Elices, M., 1998. Silkworm silk as an engineering material. Journal of Applied Polymer Science 70, 2439–2447, Blackledge, T.A., Hayashi, C.Y., 2006. Silken toolkits: biomechanics of silk fibers spun by the orb web spider Argiope argentata (Fabricius 1775). Journal of Experimental Biology 209, 2452–2461, Vogel, S., 2013. Comparative Biomechanics: Life's Physical World, second ed. Princeton University Press, Princeton, New Jersey, 628 pp.

nature: when body parts fail, the results can range from mildly annoying to life threatening. When a human breaks a fingernail or a shark breaks a tooth, they experience a mild, temporary reduction in function; the fingernail will grow back, the tooth will be replaced. In contrast, if a zebra or a jaguar breaks a leg in a fall, the consequences may be fatal: the zebra might not be able to escape from predators, the jaguar might not be able to hunt for prey. Thus, in evolution as in engineering, avoiding catastrophic failure is at a premium.

2.3.1 Fracture Mechanics: All About Cracks

Many materials fail in tension at lower stresses than they fail in compression. Moreover, the conceptual basis for failure in tension is more straightforward and historically better understood than other failure modes. We will look at tensile failure in detail because even complex structures loaded in modes other than pure tension may still begin failing in tension (as we will see in Section 2.4).

Knowing that solids are made of atoms, we might assume that failing in tension could simply be a matter of pulling on the solid hard enough to break the bonds between those atoms. In fact, from their chemical properties, we know how much energy is stored in many chemical bonds, and finding the amount of energy needed to break those bonds is a straightforward calculation. Gordon (1976, pp. 70–72) gives an example of such a calculation for steel and points out that the strength calculated on the basis of breaking chemical bonds is over 80 times greater than the experimentally measured tensile strength. If pulling apart the atoms is not what limits breaking strength, then what sets the limit?

A major reason why solids fail at much lower stresses than predicted by chemistry is that local stresses can be orders of magnitude higher than the average stress. Localized stress increases are usually caused by flaws or voids. In an isotropic material with no flaws, force trajectories—lines of equal force—through a solid in tension will run straight from end to end, all in parallel (Fig. 2.10A). If the object has a notch, the force trajectories are pinched together as they pass the notch, meaning that the same total force is squeezed across a smaller area, so the stress increases. When a notch has a sharp tip, the force trajectories get squeezed very tightly near the tip, leading to a great increase in stress right at the tip (Fig. 2.10B). Engineers call notches with sharp tips, or any opening with sharp corners, *stress raisers*, and they are to be avoided in load-bearing structures. Thus, ships are designed with round portholes and hatches with round corners, and modern airplanes always have doors and windows with rounded corners.

Moreover, the sharper the notch or crack, the greater the local stress. For an isotropic material, if the stress far from the crack is σ_0, the stress at the very tip of the crack, σ_t, is given by

(A)

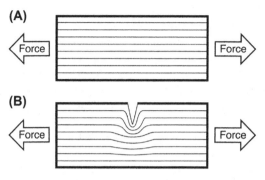

(B)

FIGURE 2.10 Force trajectories in a solid object in tension. (A) Force trajectories are parallel in a uniform material in pure tension, giving constant stress across the material. (B) Force trajectories pinch together at the end of a notch, locally increasing stress. *From Alexander, D.E., 2016. The biomechanics of solids and fluids: the physics of life. European Journal of Physics 37, 053001. Artist: Sara Taliaferro,* © *European Physical Society. Reproduced by permission of IOP Publishing. All rights reserved.*

$$\sigma_t = \sigma_0 \left(1 + 2\sqrt{\frac{l}{r}} \right) \qquad (2.4)$$

where l is crack length and r is the radius of the crack tip (Gordon, 1978, p. 67).[h] For semicircular notches or circular voids, σ_t can never be larger than three times σ_0, hence circular portholes. For narrow cracks with very sharp tips, however, where r can approach atomic diameters, σ_t can easily become orders of magnitude larger than σ_0. Seen in that light, engineers in the early 20th century began to wonder how any tensile structure managed to avoid failure at all, given that all real structures will have various flaws and many have actual cracks.

Stress concentrations at cracks and flaws seem to matter a lot more for some materials than others. A scratch in a brittle material (high stiffness, low toughness) can cause it to crack right across and fail at a relatively low stress. In fact, humans take great advantage of this property. If we score—make a straight scratch on—a piece of plate glass or ceramic tile, the glass or tile can easily be snapped in two along the score by putting the material in tension. As we will see in Section 2.4, bending an object places part in tension and part in compression, so the glass plate can easily be broken along the score by aligning the score mark with the edge of a table, clamping the part on the table, and pushing down on the overhanging part. Startlingly little force is needed, and a straight score gives a very clean break. This procedure does not work as well for other materials. Plexiglas (acrylic) plates can be scored and broken like glass, but Plexiglas requires a much deeper score, and even then getting a

---

h. This is actually the Inglis formula for an elliptical notch, but if we consider a crack to be approximated by an elliptical notch with a large semimajor axis and an exceedingly small semiminor axis, this formula gives a reasonable first approximation for an isotropic material.

clean break requires considerable skill and practice. Some other rigid materials are essentially unaffected by scoring. Scratching a thin steel plate causes no noticeable ease in fracturing it, and using the same procedure used to break glass on such a plate will likely only result in a single, slightly bent steel plate. The reason a scratch quickly propagates to failure in a piece of glass, but does not grow at all in a piece of steel, has to do with energy as opposed to stress.

Although his work took a while to become accepted, Alan A. Griffith showed theoretically that crack propagation should depend at least as much on energy as on stress (Gordon, 1976, pp. 67–71). He showed that a crack will only grow (propagate) if more energy is being released than is being used to form new crack surface. The release of energy comes from the reduction in stress as a crack gapes in tension (Fig. 2.11). The crack will only grow if the energy released by strain relief is equal to or greater than the energy required to form new crack surface area. The energy needed to form new crack surface area is W_f, the work of fracture per unit surface area (normally just referred to as the "work of fracture"). The energy released is the strain energy, W_s, and because this energy comes from reducing stress over a volume of material (Fig. 2.11), it is energy per unit volume. The ratio of W_f and W_s will be a length, and the *critical* or *Griffith crack length*, l_c is given by (Vogel, 2013, p. 332):

$$l_c = \frac{W_f}{\pi W_s}.$$
(2.5)

When the critical crack length of a material is exceeded, positive feedback occurs and the crack extends suddenly and violently, causing immediate catastrophic failure. Note that the critical crack length will be different for different materials and different for every different stress value within a given material. Brittle materials will have relatively short critical crack lengths (recall scoring glass), and tough materials will have much longer critical crack lengths. Engineering texts usually give a slightly more detailed equation for critical crack length for Hookean materials ($l_c = 2W_f/\pi\sigma\varepsilon$), but few bio-materials are Hookean, so such an equation greatly underestimates l_c for biological materials, or indeed many tough materials.

FIGURE 2.11 Tension causes cracks to gape slightly, which releases some tension from the region near the crack (shaded). *Artist: Sara Taliaferro.*

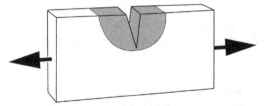

The work of fracture of a solid object contributes directly to its toughness. If a solid material requires a lot of work to produce a unit area of new fracture surface, it will be tougher than a material that takes little work to form the same unit area of new fracture surface. In fact, this difference is visibly apparent on fractured surfaces. Brittle materials tend to have smooth fracture surfaces; their low W_f means that cracks propagate quickly and cleanly straight across. Tough materials, however, usually leave very rough or ragged fracture surfaces. In energy terms, lots more surface area must be formed for cracks to propagate, which is another way of saying that a tough material forces cracks to constantly zig and zag and change directions to grow. Any features of a material that redirect cracks or increase the area of fracture surfaces (e.g., voids, embedded fibers) will therefore increase its toughness.

2.3.2 How to Stop Crack Growth

Both the material properties and the geometric properties of a solid affect its susceptibility to Griffith crack propagation. The fact that tough materials have longer critical crack lengths is intuitively reasonable, but how can shape matter? Consider glass: a solid rod or plate of glass will fail at much lower stress than a bundle of fine glass fibers of the same total cross-sectional area. Indeed, if the fibers are hair-fine, they may not fail until nearly the stress predicted by chemical bond strength. For the solid rod, failure occurs when the stress increases so that the critical crack length equals the length of the largest flaw in the rod. In a bundle of fibers, their diameters may be narrower than the critical crack length at which the rod fails. The fibers are so narrow that any flaws or impurities or scratches will simply be much smaller than the fiber diameter, so they are less likely to reach critical crack length until much higher stresses occur (Wainwright et al., 1976, p. 22). Moreover, even if a few fibers do fail due to flaws, a crack from one failed fiber cannot propagate into other fibers, so a bundle of fibers generally does not fail suddenly or catastrophically. Thus, tension-resisting structures such as tendons and cables are usually arranged as bundles of very thin fibers rather than as solid rods.

Tough materials resist crack propagation in different ways. Tough isotropic materials, such as structural metals, resist crack formation by yielding. At the tip of the crack in such a metal, the material near the crack tip may yield and stretch a bit, which absorbs some of the work of strain relief without actually fracturing. Another way to inhibit crack formation is for the material to have round or oval voids. When a crack tip hits one of these voids, the stress concentration near the tip is greatly reduced, stopping crack elongation (Fig. 2.12). Such voids may slightly reduce the material's cross-sectional area and increase the average stress, but if the voids inhibit crack extension, they can greatly increase the material's actual strength. Various mineralized structures in animals (including some kinds of bone) have voids, as do some kinds of wood.

A very effective way to limit crack propagation is to use a composite material. A composite material is one with two or more components of different stiffness. The classic engineering example is fiberglass, a composite of very fine glass fibers embedded in a resin[i] matrix. If a crack passes through a fiber, it runs into the resin, which, being less stiff, may yield a little and slow the crack. If that does not stop crack propagation, the crack may stop at the next fiber it runs into because the matrix, being less stiff than the fiber, will turn the crack to run parallel with the fiber so that it is no longer perpendicular to the stress. The most effective composites have very fine fibers, which reduces their critical crack length and provides lots of fiber-to-matrix interface area to intercept cracks. Such composites force cracks to follow a very torturous path, requiring very large amounts of energy to form new fracture surface, and this new surface will be extremely ragged and rough. While these composites will be a bit less stiff than their stiffest component, they will be much tougher than either component alone.

One of the reasons engineers make such great use of metals is their ductility, i.e., their ability to absorb a lot of energy by deforming permanently rather than fracturing. Ductility gives metals such as mild steel and aluminum alloys great toughness and makes them well suited to resisting tension. Masonry—brick, stone, unreinforced concrete—and glass have shockingly low resistance to tension. Slight flaws can easily meet or exceed the critical crack length under very modest tension. A great deal of the structure of a Gothic cathedral that might superficially appear to be ornamental—flying buttresses or finials—is actually crucial to insuring that walls only experience compression, never tension. Reinforced concrete is actually a composite, where slender, embedded steel rods resist tensile loads, while the concrete bears compressive loads.

---

i. Resin is a kind of "plastic" in the everyday sense, usually polyester or epoxy in the case of fiberglass.

Organisms have not evolved the ability to use metals in their metallic form, but they make abundant use of composites. For example, in mammals, both ropelike tendons and stiffer (but still flexible) cartilage consist of collagen fibers embedded in a protein gel matrix. The collagen in a tendon is arranged in parallel bundles, and the protein matrix is a soft gel with high water content. The collagen in cartilage is arranged in a more random, 3D, feltlike array embedded in a firmer matrix with more protein and less water. Tendon is superb at resisting tension but is too flexible to offer any resistance to compression. Cartilage, in contrast, is not especially good at resisting tension, but is so good at resisting compression that it functions as a shock absorber in many mammalian joints (such as our knees).

Wood, a composite of cellulose fibers (actually complex, multilayered tubes of helical cellulose molecules) in a protein matrix, illustrates yet another property of composites. The cellulose fibers in wood are mostly parallel, which maximize their tendency to blunt cracks perpendicular to the fibers. If the loading direction is reasonably predictable, orienting the fibers in a composite can greatly increase toughness in a particular direction, although at the cost of lower toughness in other directions. As we saw earlier, wood is very difficult to fracture across the grain—perpendicular to the fibers—due to the toughness of the cellulose fibers, but often an order of magnitude easier to fracture with the grain—parallel to the fibers.

Lacking metals, organisms often use mineral crystals such as calcite or hydroxyapatite to stiffen rigid supporting structures. The pure mineral, however, would be impractically brittle and prone to fracture. Plants and animals have gotten around this limitation in at least two ways. One is to make the mineralized structure "holey," or full of tiny, crack-blunting voids (Fig. 2.12). Spongy bone, the arrangement of bone tissue inside many mammalian bones (e.g., inside human skull roof bones and ribs), gets its name because the mineralized component forms an open, 3D meshwork of tiny struts, giving it a spongelike appearance. Echinoderms, such as starfish, sand dollars, and sea urchins, have a mineralized internal framework made of elements called ossicles, and these ossicles have a foamy or spongy appearance under a microscope. They are apparently adapted to limit crack propagation for increasing toughness.

Another way to reduce the brittleness of mineralized material is to use crystals as one component of a composite. Mollusk shells are made of calcium carbonate—the main component of limestone—but snail and clam shells are much harder to break than an equivalent thickness of pure calcium carbonate. The mineral crystals are embedded in a tough protein matrix, and such shells are much less brittle than they might appear. For example, in the 19th century near where I live, a thriving cottage industry of cutting buttons from the shells of the local freshwater clams arose. Pure calcium carbonate would be impractically brittle to cut or drill and would probably make fragile buttons, but mollusk shells yielded practical, commercially successful buttons. Vertebrate bone tissue, in addition to forming spongy bone, can be formed into the solid-appearing compact or cortical bone. Although compact bone is hard and stiff, it is not simply pure hydroxyapatite, but a complex arrangement of hydroxyapatite crystals surrounded by collagen fibers and embedded in protein, making it tough even by engineering standards. Similarly, the enamel and dentine of our teeth consist of mineral crystals embedded in protein, giving them the hardness needed for chewing food, yet the toughness to stand up to heavy daily loading for several decades.

Some organisms have even evolved nature's answer to fiberglass. Many species of sponges secrete tiny, sand grain–sized calcium carbonate or silica structures called spicules. Embedded in the sponge's tissues, such spicules deter predators and provide a small increase in stiffness. Because of their small size, the spicules are very resistant to fracturing. Some deepwater sponges, called glass sponges or hexinactinellids, have evolved an elaborate modification of this pattern: rather than tiny particulate spicules, these sponges secrete very long, thin, fiberlike spicules of silica (glass) woven into complex arrays to form a springy, stiff but flexible internal skeleton. The best studied of these, *Euplectella aspergillum*, is commonly called Venus's flower basket, and its skeleton is made up of tiny rods and threads of silica formed into an elegant, precise, basket-like arrangement. Moreover, even though the individual rods are quite narrow ($10-50 \, \mu m$), each rod is made up of many concentric, alternating layers of silica and protein, greatly increasing its toughness (Woesz et al., 2006; Monn et al., 2015). Thus, these animals have evolved a scheme to produce a tough, flexible skeleton from that most brittle of materials, glass.

2.4 STRUCTURES

2.4.1 The Engineering Categories

Engineering practice organizes solid mechanics into a hierarchy from materials to simple structures to structural systems, the higher levels being composed of elements from lower levels. As others have pointed out, organisms gain no functional benefit from organizing their structures in the engineering hierarchy (Vogel, 2013, p. 287). Nevertheless, the hierarchy gives us a starting point for analyzing more complex biological structures, as well as providing many useful concepts, such as bending and buckling.

Up to this point, we have been looking at materials. When loaded in tension, only the cross-sectional area matters, the shape or arrangement of the material has no effect (aside from fiber orientation in composites). As we shift our focus to structures, we will see types of loading where shape and orientation matter greatly; the same cross-sectional area of material can fare much better or worse, depending on how it is arranged.

The most common simple structures encountered in engineering are beams, columns, and shells. For each of these, its response to loading depends on both what it is made of—its material—and how the material is arranged—its shape.

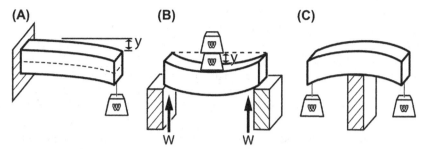

FIGURE 2.13 Deflection of loaded beams. (A) A cantilever loaded at the free end, deflected a distance y by load *W*; dashed line indicates neutral surface. (B) A simply supported beam loaded in the middle, deflected in the middle by a distance y. The force that must be supported at each end is half the load (assuming the beam's weight is negligible). (C) The loading on a simply supported beam can be represented as an upside-down version of two cantilevers back-to-back. *Artist: Sara Taliaferro.*

2.4.2 Elongate Structures: Beams

Engineers classify elongated supporting structures as *beams* if they are loaded mainly perpendicular to their long axis and *columns* if they are mainly loaded parallel to their long axis. By convention, beams are analyzed as if they were horizontal, and columns, as if vertical. Elongate animal or plant bodies or body parts, however, can function as biological beams in any orientation. Moreover, the same organismal structure can function as a beam or as a column, or even both simultaneously, depending on the situation, e.g., tree trunks resisting wind versus supporting the weight of their crown.

Beams can be *cantilevers*, with one end firmly attached and the other end free; or they can be *simply supported beams*, resting on a support at each end (Fig. 2.13). ("Fixed beams" with immovable, rigid attachments at both ends seem to be absent in nature.) Because the load on a beam is perpendicular to its long axis, beams deflect by bending (Fig. 2.13). How beams resist bending depends on both the material and the shape of the beam.

The stresses in a bent beam are very different from those of an object in pure tension or compression. The part of the beam on the outside of the bend (the top of the cantilever or the bottom of the simply supported beam in Fig. 2.13) experiences tension and elongation strain. Conversely, the surface on the inside of the bend experiences compression and shortening strain. The tension or compression is greatest at the surface and decreases toward the middle. Somewhere in the middle of the beam is a surface where tension and compression both drop to zero, this is the neutral surface.[j] In a beam of isotropic material of symmetrical cross section, the neutral surface will be in the exact center of the thickness. All beams will have a neutral surface, but if

---

j. Because engineered structures often have symmetrical shapes and small deflections, the neutral surface is sometimes referred to as the "neutral plane."

the material is anisotropic or the cross section is irregular (or both), the surface will not necessarily be in the exact middle. (The neutral surface experiences no tension or compression in bending and is stress free in simply supported beams; in cantilevers, however, it experiences a small amount of shear, normally negligible compared to the maximum tension and compression.)

If tensile and compressive stress increases with distance away from the neutral surface, then the farther a bit of beam is from the neutral surface, the more it is strained. Thus, a major concept of beam theory is that material farther from the neutral surface plays a larger role in resisting bending. The factor that determines how much a given beam deflects under a particular load is the *flexural stiffness*. Flexural stiffness is the product of the modulus of elasticity, *E*—a material property—and the second moment of the beam's cross-sectional area, *I*—a geometric property. (The second moment of area is sometimes called the area's "moment of inertia," but this moment has nothing to do with mass, so calling it a moment of "inertia" is misleading and should be avoided.) The second moment of area is a measure of the distribution of area with respect to the neutral surface; see Fig. 2.14 for formulas of *I* for some simple shapes. Since material farther from the neutral surface carries more of the load, *I* helps define how the load is distributed across the cross section. The flexural stiffness does not have its own symbol, but instead is simply referred to as "*EI*."

In typical engineering situations, both *E* and *I* are known, so calculating *EI* is simple. In biological situations, however, the cross section is rarely simple enough to allow easy calculation of *I* (and may vary along the beam's length), and *E* may not be known for a particular biomaterial. In these cases,

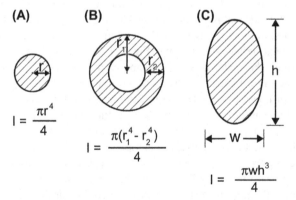

(A)

$$I = \frac{\pi r^4}{4}$$

(B)

$$I = \frac{\pi(r_1^4 - r_2^4)}{4}$$

(C)

$$I = \frac{\pi w h^3}{4}$$

FIGURE 2.14 Examples of second moment of area for beams of simple cross section. (A) Solid circular cross section. (B) Hollow cylinder of outside radius r_1 and inside radius r_2. (C) Solid elliptical cross section of major axis *h* and minor axis *w*. *Modified from Alexander, D.E., 2016. The biomechanics of solids and fluids: the physics of life. European Journal of Physics 37, 053001. Artist: Sara Taliaferro,* © *European Physical Society. Reproduced by permission of IOP Publishing. All rights reserved.*

determining *EI* experimentally as a single composite variable may be the most practical approach. The deflection, *y*, at the end of a cantilever of isotropic material and constant cross section is

$$y = \frac{FL}{3EI} \tag{2.6}$$

where *F* is the applied force (load) and *L* is the beam's length. If we want to know the flexural stiffness of a biological beam, such as a corn stalk or the wing bone of a bat, we could test it as a cantilever and apply Eq. (2.6). Firmly attaching one end of a structure of biological material can be quite challenging, so a three-point bending test is often more convenient (Fig. 2.13B). For a simply supported beam (with the same material and symmetry conditions as above) in three-point bending, with the force applied in the exact center between the supports, the formula for deflection can be rearranged to give *EI*:

$$y = \frac{FL^3}{48EI} \quad \text{or} \quad EI = \frac{FL^3}{48y}. \tag{2.7}$$

Conceptually, as Gordon (1976, p. 58) points out, a simply supported beam is equivalent to two cantilevers upside down and back to back (Fig. 2.13C). Looking at beams this way clearly shows how the greatest stress in a cantilever is at the base where it is attached, whereas the greatest stress in a simply supported beam is in the middle. Thus, cantilevers are most likely to fail near the base, but simply supported beams are most likely to fail in the middle.

If a beam is loaded in such a way that the load tends to turn or twist the beam about its long axis, the beam is loaded in *torsion*. A load that produces torsional stresses and strains is usually a *torque* or *moment* (this "moment" is technically the *moment of force*, which is rarely spelled out, and not to be confused with the moment of inertia or second moment of area). Such a torque consists of a force, *F*, times a *moment arm*, *r*. The force is applied some distance away from the beam's axis, and the distance from the beam's axis to the point of application of the force is the moment arm (Fig. 2.15). When you

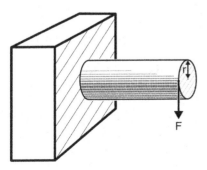

FIGURE 2.15 Terminology for cylindrical cantilever of radius *r* in torsion. Force *F* is applied tangent to the surface at the end of the beam. *Redrawn from Alexander, D.E., 2016. The biomechanics of solids and fluids: the physics of life. European Journal of Physics 37, 053001. Artist: Sara Taliaferro,* © *European Physical Society. Reproduced by permission of IOP Publishing. All rights reserved.*

turn a bicycle pedal or a crank handle, you are applying a torque or moment to a shaft. In the SI system, torques are given in newton-meter (Nm).

The stresses in a beam in torsion are complex—the beam experiences shear parallel to the outer surface, tension at 45 degrees to the long axis, and compression perpendicular to the outer surface. In fact, we routinely make use of this torsional compression when we wring a wet fabric item like a towel to squeeze out excess water.

The resistance of a beam to torsional deflection is its *torsional stiffness*, given by the shear modulus, G (Section 2.2.3), times the polar second moment of area, J. The polar second moment of area is similar to I, but describes the distribution of area relative to a neutral axis rather than a neutral plane. The torsional stiffness, GJ, like the flexural stiffness, consists of a material property multiplied by a shape property. The value of G can be calculated from E for many well-behaved engineering materials, but biological materials are generally anisotropic and non-Hookean, so such standard engineering formulas relating G and E cannot be used. Also, formulas exist to calculate J for simple shapes (e.g., $J = {}^1\!/_2 \pi r^4$ for circular cylinders) but biological beams tend to have irregular, complex cross sections, so biomechanics researchers can rarely calculate values of J from simple formulas. In fact, just as for flexural stiffness, biologists are seldom interested in the value of G (or J) alone. Thus, they usually measure the composite variable GJ experimentally, similarly to EI. If one end of such a beam is fixed and a torque is applied to the other end, then the torsional stiffness is given by

$$GJ = \frac{ML}{\theta} \tag{2.8}$$

where M is the moment (torque) applied, L is the beam's length, and θ is the angular deflection of the beam in radians. Measuring the torsional stiffness of a biological beam can be challenging because one end of the beam must be firmly fixed. Clamps tend to damage the sample and cause stress concentrations, so generally fixing the end into some sort of holder with epoxy or cyanoacrylate adhesive gives better results.

2.4.3 Elongate Structures: Columns

If a beamlike structure is loaded on its ends in pure compression, we consider it a column, and such a structure deflects differently from a beam. When short, squat columns are compressed, they act much like tensile structures in reverse: the structure shortens with a stress—strain relationship that may have the same or very similar E, and they fail by crushing or by cracks allowing shearing. Narrow columns, however, are much more likely to fail at lower stresses due to sideways deflections called *buckling*. Columns can experience two different kinds of buckling. *Euler buckling* involves the whole column, and in the simplest form, the middle (along the length) of the column deflects to one side as the ends get closer together, forming one, continuous, smooth, arc-shaped curve. (Depending on conditions, an S-shaped or even more complex curved pattern may also develop.) The wall of the column on the outside of the bend is in tension, whereas the wall of the column on the inside of the curve is in compression (Fig. 2.16). Such a beam will normally fail when tensile stress in the convex region becomes large enough to propagate cracks.

The other form of buckling is *local buckling*. This form mainly applies to hollow, thin-walled columns. It occurs when a kink or crease appears along the wall, often where compressive stress is high. Such a local distortion usually happens at the site of some minor flaw, in an analogous way to a flaw triggering Griffith crack propagation in pure tension. Also like a propagating crack, local buckling can relieve stress and liberate strain energy, so that the kink or crease rapidly propagates around and along the column, leading to sudden collapse.

The *critical force*, F_c, for a column is the force that is just enough to cause the column to fail by buckling. For Euler buckling, F_c is given by

$$F_c = \frac{n\pi^2 EI}{L^2} \tag{2.9}$$

where n is a coefficient with a range from 1 to 4 that depends on how the ends of the column are secured; e.g., if both ends can bend or twist freely, $n = 1$, whereas if both ends are firmly fixed, $n = 4$. If one end is firmly fixed (such as a firmly rooted tree trunk), then n is 2 (Beer and Johnston, 1981, p. 529). Eq. (2.9) shows that for Euler buckling, F_c is proportional to flexural stiffness and inversely proportional to (length)^2, meaning that geometry plays a large role in resistance to buckling.

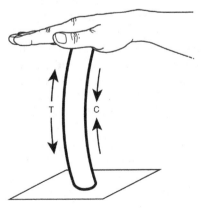

FIGURE 2.16 Euler buckling of a cylindrical column. As the column deflects laterally, the material on the outside curvature experiences tension (T) and the material on the inside curvature experiences compression (C). *Artist: Sara Taliaferro.*

Because local buckling is partly a function of flaws in the material, it is not well suited to a theoretical analysis, so engineers use a variety of semi-empirical relationships to analyze it. For example, Ennos (2012) gives

$$F_c = k\pi t^2 E \qquad (2.10)$$

where t is the hollow column's wall thickness and k is a coefficient ranging from 0.5 to 0.8 that accounts for the likelihood of flaws and impurities.

Elongate biological supports—leg bones, plant stems, tree trunks, sea anemone columns—can function as both beams and columns, often simultaneously. Eqs. (2.7) and (2.8) for beams and Eqs. (2.9) and (2.10) for columns assume circular struts of constant cross section and isotropic material. Very few biological columns fit that description. Nevertheless, these equations offer an informative starting point because they suggest geometries that should be less prone to failure; struts with near-circular, hollow cross sections are fairly common in biology. Even though the equations may not be able to give exact values for deflections or failure loads, they illustrate general design principles and suggest what factors may or may not be mechanically significant in an organism's structure. For example, Parle et al. (2016) showed that buckling failure of some insect leg segments could be predicted analytically, whereas other insect leg segments did not fit the predictions based on standard equations.

2.4.4 Shells

In engineering terms, a *shell* is a hollow structure with walls of generally spherical or ellipsoidal shape. Engineers focus mainly on the properties of hemispherical domes, whereas biological shells include both partial spheres such as the braincase of our skulls and complete spheres or ellipsoids such as eggshells and sea urchin skeletons ("tests"). Such biological shells can experience loading in various ways, including from internal or external pressure, from squeezing or crushing, or from piercing. Squeezing loads affect a shell in the same way that gravitational self-loading affects a section of the wall of a building with walls in the shape of a hemispherical dome. A load on the top of such a dome produces meridional compression—radiating from the top down the sides of the dome, like meridians of longitude—and parallel tension, perpendicular to the meridional compression, oriented like the equator and lines of latitude. For biological shells, the meridional compressions are more likely to be limiting than the parallel tensions. For this kind of loading, Vogel (2013, p. 386) shows that meridional stress will vary directly with length, so larger shells should be disproportionately thick. Bird eggshells do increase in thickness in proportion to (length)$^{1.38}$, but this is considerably less than the (length)^2 proportionality needed to keep stress constant (Ar et al., 1979). In fact, larger eggshells seem to be made of slightly stronger material than smaller eggshells, but large eggs still failed at proportionately lower stresses than small eggs. For a detailed discussion of crushing and piercing resistance in biological shells, see Vogel (2013, pp. 384–388).

The mechanics of pressurized shells and domes is very similar to that of pressurized cylinders, so we will defer that discussion until Chapter 5.

2.4.5 Examples of Biological Structures

Some elongated biological supports function as beams, some function as columns, and many do both. One example of a biological beam is a horizontal tree branch. Wood of such a branch is made up of longitudinally arranged fibers formed by the cell walls of stacks of long, narrow cells. The cell walls, in turn, are made of complex layers of cellulose fibers—a polymer of glucose—in an organic matrix stiffened by a phenolic compound called lignin. Because of this anisotropic arrangement, the tensile strength of a branch perpendicular to the grain (fiber direction) is much lower than parallel to the grain. Thus, when a small horizontal branch is loaded enough to cause a sharp curve, it breaks with a characteristic "greenstick" fracture, the wood on the outside of the curve fails in tension, and a crack propagates to the middle of the branch, where it diverts longitudinally to form the classic lengthwise split (Fig. 2.17). This leaves about half of the branch's diameter unbroken, which forms a flexible hinge and is nearly impossible to break using only further bending.

Some of the segments of arthropod legs can function as beams. Many spiders, and a few insects, hold certain leg segments horizontally when walking on level surfaces (Fig. 2.18). These segments are typically near circular in cross section with thin walls and a hollow interior containing muscles and hemolymph (blood). The walls of these leg segments are extremely thin compared to the relative wall thickness of hollow vertebrate bones, but the spider's very low mass means that these apparently spindly tubes have enough material far enough from the neutral axis to give them the flexural and torsional strength to avoid damage from bending or twisting.

Wings that support a flying animal's body in flight must resist bending from base to tip to carry flight loads, and we will see in Chapter 3 that such

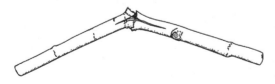

FIGURE 2.17 A characteristic greenstick fracture. When a stick of green wood bends, it fractures transversely on the tension side and splits longitudinally down the middle on both sides of the fracture. This is a characteristic of stiff but slightly flexible solids such as living wood and long bones of human children. *Artist: Sara Taliaferro.*

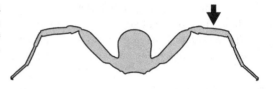

FIGURE 2.18 Outline of a transverse section through a spider. Arrow indicates a segment of the leg that acts as a horizontal beam. *Artist: Sara Taliaferro.*

wings must also resist (or control) torsion as well. Bird wing bones, for instance, are at the leading edge (front) of the wing, and since the center of lift is about one quarter of the way back from the leading edge, these bones must resist both bending and torsion. The wing bones are hollow cylinders; although the walls look thin, they have a high enough I to resist bending and J to resist twisting under normal flight loads. In contrast, the primary or pinion feathers have their shafts placed so as to limit torsional loading. The primary feathers make up the bird's wing tip, and in most birds they can be spread out so that each feather functions as a tiny, individual wing. Each feather's shaft is located about one-fourth of the way back from the leading edge, where the lift is centered. Thus, the feather shaft must resist bending but not much torsion, and so feather shafts, although hollow, are often more D-shaped, or even U-shaped, than circular.

Insect wings are based on a completely different design from birds, with longitudinal struts or "veins" supporting a thin, flexible membrane. For such minute structures, insect wings incorporate several clever design features. They are not simply flat plates; flat plates would be aerodynamically ineffi-cient as well as prone to bending and twisting. Insect wings are corrugated—arranged like flattened pleats—with a stiff vein at each peak and valley. This increases the wing's I perpendicular to the wing span and makes it resistant to bending from base to tip. Wings work best if they are cambered (convex upward from front to back, see Section 3.4), and because of the way the wing veins (and corrugations) fan out from the leading edge, twisting the leading edge vein automatically cambers the wing (Ennos, 1988).

Many plants have evolved elongated support structures that resist bending much more strongly than torsion. Vogel (1992, 1995) noticed this pattern and suggested the *twist-to-bend ratio*[k] to quantify it. The twist-to-bend ratio, T/B, is given by

$$T/B = \frac{EI}{GJ}. \tag{2.11}$$

Note that EI is the resistance to bending and GJ is the resistance to twisting. Vogel's insight was that $(GJ)^{-1}$ is the torsional compliance (ease of

---

k. Vogel actually uses the term "twistiness-to-bendiness ratio" in his books (e.g., Vogel, 2013, pp. 381–384).

twisting) and $(EI)^{-1}$ is the bending compliance (ease of bending), so T/B is actually the ratio of $(GJ)^{-1}$ to $(EI)^{-1}$, hence twist-to-bend ratio.

Many plants reorient to reduce drag in the wind. Daffodils carry horizontally oriented flowers on long stems to attract particular pollinators. In wind, the stem twists to allow the flower to trail downwind, which reduces the aerodynamic load on the flower and stem. The daffodil's stem has a T/B in the range of 7−13 (Etnier and Vogel, 2000; Ennos, 2012). For comparison, a spring steel rod has a T/B of 1.3, a maple leaf petiole (leaf stem) 2.3, and a mammal leg bone of 2.9 (Vogel, 2013, p. 383). In contrast, tree trunks, despite an appearance of great rigidity, can have T/B values > 7 (Vogel, 1995), and other plants appear even more specialized to allow reorientation. For example, the T/B of the stems of a type of sedge average approximately 40 (Ennos, 1993), and the petioles of banana leaves average nearly 70 (Ennos et al., 2000). Wind-pollinated sedges thus avoid self-pollination, and bananas reduce the wind loads on their enormous leaves in violent tropical storms. Vogel (1992, 2009) attributes these high T/B values to the structures' decidedly noncircular cross sections: flattened, triangular, V-shaped, or U-shaped. Ennos noted that cross section alone is not sufficient to explain really high T/B values. He pointed out isolated bundles of woody or other supporting tissue in such stems, giving them very low resistance to shear and thus facilitating twisting, while still providing some bending resistance (Ennos, 2012, p. 164).

Some biological support structures function as columns, resisting gravitational compression. Our leg bones, and the leg bones of most large mammals, function as columns when we stand still or when we land from a jump. Tree trunks and most plant stems function as columns when they support their crown or leaves in still air. Many plant stems have hollow, circular cross sections, e.g., those of dandelions, tulips, wheat, or bamboo. A hollow cylinder is excellent for resisting bending, torsion, and Euler buckling, but can be prone to local buckling. Engineers often solve this problem with bulkheads— transverse walls across the cylinder's diameter—or stringers—lengthwise thickenings or reinforcements on the inside or outside of the column's walls (Fig. 2.19). Bulkheads for reinforcing columns may be most familiar as the septa or nodes in bamboo, but similar partitions occur in other hollow plant stems. Their function is to prevent "ovalization"—allowing the diameter to decrease in one direction and increase in another—or crimping that can initiate local buckling. Surprisingly, biological bulkheads do not maintain circularity by resisting compression. All of those studied so far work in tension, the same way as bicycle spokes (Ennos et al., 2000; Ennos, 2012, p. 132); in other words, they resist increases in diameter using tension rather than resisting decreases in diameter using compression. Such tension-resisting bulkheads are much lighter and cheaper to make than compression-resisting bulkheads.

Many biological struts function as both beams and columns. The long bones of the legs of a galloping horse, for example, function as a column when the foot hits the ground and the leg supports the horse's weight, and a beam

FIGURE 2.19 Reinforcements against buckling in hollow cylinders. (A) Transverse partitions (bulkheads or septa). (B) Longitudinal stiffeners (stringers); shown here on the inside but they work equally well on the outside. *Artist: Sara Taliaferro.*

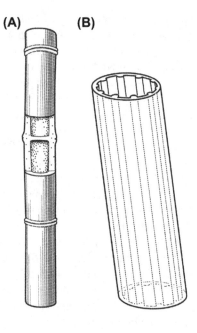

(A)   **(B)**

when the horse's muscles apply the forces and torques needed to power its movements. Trunks of large trees, as mentioned earlier, function as columns in still air, but also function as beams in the wind when aerodynamic forces on the crown apply bending and twisting loads to the trunk. Wood is often much stronger in tension than compression, providing trees with a dilemma: when they bend in a wind, wood on the downwind (compression) side could be crushed, forming cracks that would quickly fail in tension if the wind reversed direction. James Gordon, an insightful structural engineer, pointed out that living tree trunks should be prestressed in tension (Gordon, 1978, pp. 280–281), and much research has shown this to be true (Jeronimidis, 1980). The wood in a tree trunk grows in such a way that the outer layers are maintained in tension, which puts the central heartwood in compression. When wind bends a tree trunk, on the downwind side compression at first merely reduced the pre-tension, and even with greater bending, the compression is always much lower than the tension on the upwind side. Because the wood can withstand much higher tensions than compressions, such pretensioning allows a tree trunk to bend almost twice as far without failing than if it were not pretensioned (Ennos, 2012, p. 155).

The most familiar biological shells are probably those that fit the everyday meaning of shells: the protective shells of mollusks such as snails and clams. Mollusk shells are mostly made of the mineral calcium carbonate, with a relatively small amount of protein matrix, 4% or less. The calcium carbonate can be in at least two different crystalline forms, which can in turn be organized into several different arrangements. For example, mother of pearl, or *nacre*, is formed from flat aragonite crystals arranged into many sheetlike layers parallel to the shell's surface (Ennos, 2012, pp. 90−91). A typical shell is composed of two or more mineralized layers, each layer of different crystal arrangement with different mechanical properties. Although researchers have measured the mechanical properties of some of the crystalline arrangements (Currey, 1977, 1980b), most mollusk shells have such complex shapes that quantitative mechanical analyses of the whole shell are impractical. Moreover, defining the primary function (or functions) can be challenging. Has a particular shell evolved to resist compression, crushing, chipping, abrasion, to provide weight for stabilization, or streamlining to minimize effects of waves or currents? In the words of Wainwright et al. (1982, pp. 261−262), "Usually one has little idea and, frequently, one has the thought that the answer is: 'none of these'."

Many other types of organisms use shells (in the engineering sense). Researchers have discovered quite a bit about the mechanical properties of some, such as mammal skulls and turtle shells (Hu et al., 2011) and bird eggshells (Ar et al., 1979). Much less is known about the mechanical behavior of other common shell structures, such as insect thoraxes, crab carapaces, or sea urchin skeletons.

For readers interested in learning more about solid biomechanics, some of the books mentioned in Chapter 1, such as those by Wainwright et al. (1982), Alexander (1983) and Vincent (1990), are still excellent introductions to the foundations of the field, whereas more current books such as those by Ennos (2012) and Vogel (2009, 2013) provide a more up-to-date review of current research. Finally, no one interested in this topic should pass up James E. Gordon's *Structures, Or Why Things Don't Fall Down* (1978). This witty, erudite, eminently readable book is one of the best books on science for a general audience and certainly the clearest, most readable book on an engineering topic I have ever read.[1]

FURTHER READING

General References, Classic

Alexander, R.M., 1983. Animal Mechanics, second ed. Blackwell Scientific Publications, Oxford. 301 pp.

Gordon, J.E., 1978. Structures, or Why Things Don't Fall Down. Penguin Books, Harmondsworth, Middlesex, England, 395 pp.

---

1. *The New Science of Strong Materials, Or Why You Don't Fall Through the Floor* (Gordon, 1976), by the same author, is intended more as a textbook but is still very clearly and amusingly written and nearly as enjoyable as *Structures*.

Vincent, J.F.V., 1990. Structural Biomaterials. Princeton University Press, Princeton, New Jersey, 206 pp.

Wainwright, S.A., Biggs, W.D., Curry, J.D., Gosline, J.M., 1982. Mechanical Design in Organisms (paperback ed.). Princeton University Press, Princeton, New Jersey, 423 pp.

General References, Recent

Beer, F.P., Johnston, E.R., DeWolf, J.T., 2006. Mechanics of Materials. McGraw-Hill Higher Education, Boston, 787 pp.

Ennos, A.R., 2012. Solid Biomechanics. Princeton University Press, Princeton, New Jersey. 264 pp.

Vogel, S., 2013. Comparative Biomechanics: Life's Physical World, second ed. Princeton University Press, Princeton, New Jersey. 628 pp.

Collagen and Tendons

Alexander, R.M., Vernon, A., 1975. The mechanics of hopping by kangaroos (Macropodidae). Journal of Zoology 177, 265−303.

Biewener, A., Alexander, R.M., Heglund, N.C., 1981. Elastic energy storage in the hopping of kangaroo rats (*Dipodomys spectabilis*). Journal of Zoology 195, 369−383.

Plant Mechanics

Ennos, A.R., Spatz, H.C., Speck, T., 2000. The functional morphology of the petioles of the banana, *Musa textilis*. Journal of Experimental Botany 51, 2085−2093.

Etnier, S.A., Vogel, S., 2000. Reorientation of daffodil (*Narcissus*: Amaryllidaceae) flowers in wind: drag reduction and torsional flexibility. American Journal of Botany 87, 29−32.

Jeronimidis, G., 1980. Wood, one of nature's challenging composites. In: Vincent, J.F.V., Currey, J.D. (Eds.), The Mechanical Properties of Biological Materials, vol. 34. Cambridge University Press, Cambridge, UK, pp. 169−182.

Protein Rubbers

Bennet-Clark, H.C., Lucey, E.C.A., 1967. The jump of the flea: a study of the energetics and a model of the mechanism. Journal of Experimental Biology 47, 59−76.

Weis-Fogh, T., 1960. A rubber-like protein in insect cuticle. Journal of Experimental Biology 37, 889−907.

Shell and Bone

Currey, J.D., 1980. Mechanical properties of mollusc shell. In: Vincent, J.F.V., Currey, J.D. (Eds.), The Mechancial Properties of Biological Materials, vol. 34. Society for Experimental Biology, Cambridge, UK, pp. 75−97.

Hu, D.L., Sielert, K., Gordon, M., 2011. Turtle shell and mammal skull resistance to fracture due to predator bites and ground impact. Journal of Mechanics of Materials and Structures 6, 1197−1211.

Silk

Blackledge, T.A., Hayashi, C.Y., 2006. Silken toolkits: biomechanics of silk fibers spun by the orb web spider *Argiope argentata* (Fabricius 1775). Journal of Experimental Biology 209, 2452−2461.

Denny, M.W., 1980. Silks: their properties and functions. Symposium of the Society for Experimental Biology 34, 247−272.

Pérez-Rigueiro, J., Viney, C., Llorca, J., Elices, M., 1998. Silkworm silk as an engineering material. Journal of Applied Polymer Science 70, 2439−2447.

Chapter 3

Fluid Biomechanics

3.1 FLUID BASICS

As large, terrestrial, walking organisms, our everyday experiences do not prepare us well for understanding the physics of fluids. Swimming and flying animals, particularly the small ones, experience a world very different from ours. Many of the most basic concepts and processes—Bernoulli's equation, the no-slip condition, the Reynolds number—may seem unfamiliar or even counterintuitive at first.

3.1.1 Fluids Defined and How to View Them

The first such concept is that fluids include both liquids and gases. We can lump liquids and gases together because in all biologically relevant situations, flows of liquids and gases obey the same relationships: their flows are *incompressible* (see Box 3.1). In mechanics, the distinction between solids and fluids is that solids resist being deformed, whereas fluids resist the *rate* of deformation, not deformation itself. Because solids resist deforming, they maintain a fixed shape. Fluids do not resist deforming and do not maintain a fixed shape, and are thus capable of more or less freely flowing. What fluids do

---

BOX 3.1 Incompressible Gases

The fact that gases can easily be compressed, and liquids cannot, may matter in some situations, but this distinction plays no role in biologically relevant fluid flow patterns and flow relationships. For the compressibility to affect the flow patterns, the flow speeds must approach the speed of sound—approximately 340 m s^{-1} (760 mph) in air at sea level. At such speeds, the pressure in the flows can affect the density, hence such flows are called *compressible*. For slower flows, the flow speeds and patterns themselves are not energetic enough to compress the fluid, so such flows are *incompressible*, regardless of whether they are gases or liquids. Compressible flows add an extra layer of complexity to the analysis of flows; fortunately for us, I do not know of any biological situations where compressible flows occur, so all the flows we consider will be incompressible.

Nature's Machines. http://dx.doi.org/10.1016/B978-0-12-804404-9.00003-7

resist is rate of deforming, and in particular, rate of *shearing*. Imagine a volume of fluid as being made up of many infinitely thin layers all stacked up on top of each other. The layers can slide past each other essentially forever—this is shearing—but the faster they slide past each other, the more they resist, i.e., the fluid responds to the rate of shearing. Fluids resist faster shearing more strongly than slower shearing.

Because flowing fluids by definition involve motion, we need to consider a question of viewpoint that did not arise in our survey of solid mechanics. If we are interested in the flow patterns of a flowing fluid, we can envision the fluid moving past a stationary point in space or we can envision the movement of an infinitesimally small packet of fluid moving with the fluid flow. Engineers call the former approach an Eulerian perspective and the latter a Lagrangian perspective, and for our purposes they are interchangeable. If I want to know the resistance to flow of an object, say, a preserved fish carcass, I can tow it through still water at 2 m s^{-1} or I can hold it steady in a water tunnel (flow tank) with a water speed of 2 m s^{-1}; the forces and flow patterns will be the same. Either perspective has advantages and disadvantages for analyzing the fluid mechanics, but the key point is that as far as the forces and flow patterns are concerned, they are equivalent. This ability to change perspectives makes visualizing flows around moving animals much easier, not to mention allowing the use of wind tunnels and flow tanks to study flow patterns around body appendages that would be difficult or impossible to observe on a freely moving animal.

Given that all flows of relevance to organisms are incompressible, conservation of mass leads to a simple but powerful concept, the *continuity principle*. The continuity principle says that for any pipe with rigid walls, the volume that enters one end in a given time must be the same as the volume that comes out at the other end in that time (Vogel, 1994, p. 32; Fox and McDonald, 1998, pp. 107–108). If the fluid cannot change density and the walls do not expand or contract, the fluid has nowhere to hide, so the same amount that goes in must come out the other end. This principle can be stated quantitatively as

$$S_1 \bar{v}_1 = S_2 \bar{v}_2 \tag{3.1}$$

where S_1 is the cross-sectional area at location 1, $\bar{v}_1$ is the average flow speed at location 1, and S_2 and $\bar{v}_2$ are the corresponding area and speed at location 2; Eq. (3.1) will hold for any pair of locations anywhere in the pipe. The cross-sectional area times the average flow speed gives the volume flow rate, Q, so the continuity principle means that Q is constant anywhere in the pipe, regardless of constrictions or expansions. Thus, if the pipe gets narrower (lower S), the mean speed increases; and if the pipe widens, speed must decrease. Moreover, if the pipe branches into several smaller pipes, the sum of the Q's for all the branches must equal the Q of the original pipe. The continuity principle applies to most internal fluid transport systems (e.g., respiratory and circulatory systems), and we will see that it even applies to external flows under certain conditions.

3.1.2 Viscosity

The resistance of a fluid to shear rate is a material property called the *viscosity*. Viscosity is the form taken by friction in fluids. Intuitively, viscosity's everyday meaning is more or less the same as its technical meaning: thin, runny fluids have low viscosity, whereas thick, goopy, less freely flowing liquids have high viscosity. Quantitatively, viscosity gives the relationship between the applied shear stress and the rate of shearing for a given fluid:

$$\frac{F}{S} = \mu \frac{\gamma}{t} \qquad (3.2)$$

where F is the applied force, S is the area parallel to the force, γ is the shear angle (as in Fig. 2.5), so γ/t is the shear rate for a constant force (the shear rate is $d\gamma/dt$ for a varying force) (Shaughnessy et al., 2005, p. 16). The symbol μ is the dynamic viscosity, also called the coefficient of viscosity (in some fields, the symbol η is used instead of μ), and dynamic viscosity corresponds closely to our everyday usage of "viscosity." Note that "viscosity" used without a modifier is assumed to mean "dynamic viscosity," as opposed to other variations such as kinematic viscosity or relative viscosity.

Another way to view dynamic viscosity is to imagine two flat plates of area S separated by a layer of fluid of thickness z (Fig. 3.1). If the lower plate is stationary and the upper plate moves parallel to the lower plate at speed v, then

$$\frac{F}{S} = \frac{\mu v}{z} \text{ or } \mu = \frac{Fz}{Sv}. \qquad (3.3)$$

Note that F/S equals the shear stress, τ, in the fluid, and v/z is another way of expressing the shear rate.

Although μ is a material property, it can be significantly affected by temperature. As air temperature increases from 0 to 40°C, the viscosity of air

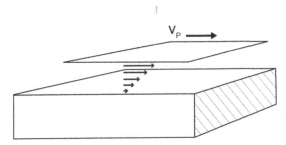

FIGURE 3.1 Shearing a fluid. The upper plate moves over a region of fluid, separating it from the immobile lower plate. The length of each arrow represents the fluid speed at that location. V_p, speed of the upper plate. *Artist: Sara Taliaferro.*

increases by about 20%. In contrast, over the same temperature range, the viscosity of freshwater actually *decreases* by over 50%. Vogel (1994, pp. 27–29), among others, has speculated about the possible biological consequences of the large change in water's viscosity over a biologically significant temperature range.

3.1.3 Drag

Drag is any force that tends to slow objects moving through a fluid or slow a fluid flowing over an object (Shaughnessy et al., 2005, p. 132). Technically, drag is a force, D, parallel to the object's motion but in the opposite direction. Drag can come from a variety of processes. For instance, in order for solid objects to move through fluids or for fluids to flow past objects, some fluid will be sheared. The viscosity of the fluid will cause some resistance to these movements. Such resistance is one form of drag.

When fluid flows around a solid obstruction, some shearing may occur just from the deflection of the flow. The lion's share of the shearing, however, comes from a key fluid property called the *no-slip condition* (see Box 3.2). In a nutshell, the fluid's velocity *at the solid's surface* must be zero relative to the solid (Shaughnessy et al., 2005, p. 336). If the fluid's speed is zero at a solid surface and the fluid is moving at some arbitrary free stream velocity, v_∞, somewhere else, then in between the location of the solid surface and the location of the free-stream speed, a velocity gradient must exist. This velocity gradient contains significant shearing, and the shearing in the velocity gradient is the main source of resistance due to viscosity, i.e., viscous drag. We will look at velocity gradients in more detail in Section 3.3; for now, the key points are that the fluid velocity smoothly approaches zero as you approach the surface and the fluid velocity is actually zero at the fluid–solid interface; note that the velocity *gradient* is not zero at the interface, due to the near-linear decrease in velocity close to the interface. Although the velocity gradient may be on the order of only millimeters or even micrometers thick, it covers

---

BOX 3.2 The No-Slip Condition

The no-slip condition says that at a fluid–solid interface, the fluid in contact with the solid surface will have zero velocity relative to the solid. In other words, an infinitesimally thin layer of fluid right in contact with the solid surface has the same velocity as the solid, as if it were sticking to the solid surface. Moreover, the flow speed does not approach zero asymptotically as distance to the solid decreases, but decreases nearly linearly close to the interface. Although counterintuitive, all biologically relevant fluids obey the no-slip condition, regardless of the nature of the solid surface (smooth or rough, hydrophobic or hydrophilic, etc.) (Vogel, 1994, p. 19).

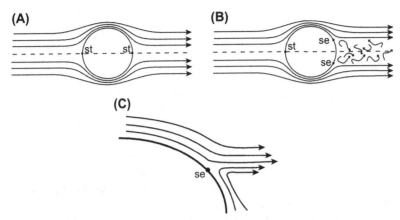

FIGURE 3.2 Fluid flowing over a cylinder (see Box 3.3). (A) In an ideal fluid (no viscosity), streamlines leaving the back are a perfect mirror image of the streamlines approaching the front; *st*, stagnation point. (B) In a real fluid with viscosity, streamlines do not get all the way around the back of the cylinder, leaving a low-pressure wake; *se*, separation point. (C) Close-up view of separation point. *From Alexander, D.E., 2016. The biomechanics of solids and fluids: the physics of life. European Journal of Physics 37, 053001, © European Physical Society. Reproduced by permission of IOP Publishing. All rights reserved.*

and affects the entire surface area of the solid exposed to the fluid flow. Viscous drag is also called "skin friction" drag because it is generated by the entire surface ("skin") in contact with the fluid (Shaughnessy et al., 2005, p. 177). Thus, all else being equal, objects with greater surface area will have greater viscous drag.

Viscosity contributes indirectly to a phenomenon that produces another form of drag, pressure drag. To understand pressure drag, it helps to consider what happens in an "ideal fluid," an imaginary material with no viscosity. (Based on Newton's analysis, later physicists found this to be a useful concept because the relevant equations had analytical solutions, unlike the more general equations that include viscosity.) Imagine a cylinder in a fluid flow perpendicular to the cylinder's long axis (Fig. 3.2A). In an ideal fluid, the flow approaches the front of the cylinder with half of the flow going over the top and half going under the bottom. On the front face of the cylinder is a point called the *stagnation point*, an infinitesimally small point where the fluid comes to a stop. Over the front half of the cylinder, the fluid accelerates as it flows along the surface to a maximum speed at the very top and bottom. Past the upper- and lower-most points of the cylinder, the flow decelerates as it flows along the back surface. The upper and lower flows are again separated by another stagnation point in the middle of the rear face,[a] and the fluid flows

---

a. From Bernoulli's equation, we can calculate that the pressure at both stagnation points equals the total of the static and dynamic pressure (see Section 3.2.1).

away behind the cylinder in the exact mirror image of the pattern of the fluid approaching the front. This analysis led early fluid theorists to what is now called d'Alembert's paradox: the pressures on the front and back are perfect mirror images and thus cancel out. The cylinder experiences no drag!

In a real fluid with actual viscosity, the flow pattern around a cylinder looks distinctly different. The flow over the front face looks about the same, but the flow over the back half is quite different. The difference is due to viscosity: as the flow accelerates on the front face of the cylinder, viscosity slightly reduces the acceleration and removes a bit of momentum from the flow. Then as the fluid decelerates on the rear face, it reaches a speed equal to the speed at which it originally approached the front face before it gets all the way around the back. That is, the fluid loses enough momentum to viscosity on the front face that it does not have enough momentum to get all the way around the back face. Instead, the flow from the top and bottom peels away from the surface at *separation points* (Fig. 3.2B and C) well before reaching a rear stagnation point. This separation forms a low-pressure wake. The difference between the high pressure on the front and the low pressure on the back represents a net force in the direction of the flow (Shaughnessy et al., 2005, p. 903). This force is called *pressure* or *form drag*, or equivalently *inertial drag*, because the fluid does not have enough inertia to flow all the way around the back of the cylinder to the rear stagnation point. So while it is true that the pressure differences on front and back are what produce this net force, the loss of momentum that leads to this difference is ultimately due to the presence of viscosity.

The magnitude of the drag force can be strongly influenced by size, so scientists have developed a nondimensional index of drag that is useful for comparisons. This dimensionless index is the drag coefficient, C_D, given by

$$C_D = \frac{D}{\frac{1}{2}\rho v^s S} \tag{3.4}$$

where ρ is fluid density, v is speed, and S is some reference area. (Note that Eq. (3.4) is often rearranged to put D on one side of the equal sign and called the "drag equation" as if it defines drag. In fact, it is not a definition of drag, but simply a statement that the drag coefficient relates the drag to a particular reference force based on the dynamic pressure.)

One confusing aspect of the drag coefficient is the choice of reference area. Engineers have commonly used four different areas. One is the *frontal area*, the projected area of the object looking at it from the direction of on-coming flow. This is normally the reference area used for nonstreamlined ("bluff") objects. Another is the *wetted area*, the total surface area in contact with the fluid (even if the fluid is air), and by convention, this area is used for streamlined objects, i.e., objects with a tapered rear shape to reduce pressure drag (described in more detail in the next section). The third is the *planform area*, the vertically projected area looking down from above, used for any wing or lift-producing object. A final possible choice is *volume*$^{2/3}$, originally used for lighter-than-air vehicles. Vogel (1994, p. 91) makes a compelling case that the last of these reference areas is particularly appropriate for organisms, given the difficulty of accurately measuring body surface area and the fact that optimizing the volume available for organs such as gonads may be under pressure by natural selection. In any case, as Vogel also points out, any published drag coefficient data that do not specify which reference area was used in the calculations are scientifically useless.

3.1.4 The Reynolds Number

Viscous drag and pressure drag scale quite differently from each other. For example, viscous drag generally varies directly with the flow speed, whereas pressure drag varies with the square of the speed. Scientists use a dimensionless ratio called the Reynolds number to give a measure of the relative importance of viscous or pressure drag in particular conditions. The Reynolds number, Re,[b] is given by

$$Re = \frac{\rho v l}{\mu} \tag{3.5}$$

where l is a reference length, usually the length of the object parallel to the flow. The Reynolds number represents the ratio of inertial forces to viscous forces (Fox and McDonald, 1998, p. 306). Thus, a low Reynolds number means viscous drag dominates, and a high Reynolds number means pressure drag dominates. In biomechanics, knowing the Reynolds number to order of magnitude, or at most, one or two significant figures, is usually sufficient precision (although certain highly technical applications may require higher precision). For biological purposes, Re values less than 1.0 are dominated by viscous drag and values greater than 10^5 are dominated by pressure drag with negligible viscous drag. Most animals (and macroscopic plants) encounter flows at Reynolds numbers between these two extremes, and because living organisms tend to have complex shapes, empirical measurements may be more informative than the Re alone in determining the significance of viscous and pressure drag.

Nevertheless, the Re can be a useful tool to compare organisms across wide size ranges. At low Reynolds numbers, the dominance of viscosity means that velocity gradients are thick, and this is the regime of small, slow-moving creatures. Conversely, at high Reynolds numbers, plants and animals are large, locomotion and flow speeds (wind, currents) are fast, and velocity gradients near surfaces are very thin. These differences have many biological consequences, affecting physiological processes such as heat and gas exchange, as well as biomechanical processes such as locomotion. For

---

b. By convention, it is "Reynolds number," not "Reynolds' number," and "Re," not "R$_e$."

example, the effectiveness of wings varies significantly with the Reynolds number, as we will see in Section 3.4.

Beyond drag comparisons, the Reynolds number is also useful for analyzing whether a flow will be laminar or turbulent. If a fluid follows smooth, predictable streamlines as it flows, it is called *laminar* (see Box 3.3). If a flow is full of chaotic swirls and eddies, it is called *turbulent*. In fact, Osborne Reynolds initially developed the index that now bears his name as a way to predict whether flows in a cylindrical pipe would be laminar or turbulent: flows with a low index are laminar, whereas flows with a high index are turbulent (Reynolds, 1883). A dye stream introduced from a narrow tube into the flow will show the difference between laminar and turbulent flows unequivocally: in laminar flow, the dye will continue down the pipe as a

BOX 3.3 Streamlines

A streamline is a line or curve in a steadily flowing fluid where the fluid velocity is always tangent to the line. A streamline could be constructed by measuring the fluid velocity, moving a tiny distance in the direction of the velocity, measuring the velocity again, moving again, and repeating until the desired length of the streamline was achieved. Another way to visualize a streamline is to imagine an infinitesimally small particle of fluid with negligible inertia being carried along with the fluid: that path of such a particle will trace a streamline.

Although cumbersome to describe, streamlines have some very important properties. One is that, since the velocity is, by definition, tangent to streamlines, fluid *cannot cross* streamlines. Because fluid cannot cross streamlines, the continuity principle holds between any pair of streamlines, i.e., the volume flow rate is constant between streamlines. Thus, in steady flows, streamlines indicate relative speeds: when streamlines pinch close together, the fluid must flow faster, just as if they formed a pipe that narrowed; when streamlines spread apart, the fluid slows. A pattern of streamlines thus functions as a sort of contour map of flow speeds. A pattern of streamlines provides a useful visual summary of a flow, as well as specific flow velocity data when fluids flow around complex objects such as organisms (Vogel, 1994, pp. 41–43).

In practice, researchers usually approximate streamlines using pathlines or streaklines. A *pathline* is a path followed by a small, neutral-density particle in the flow, usually visualized with either a continuous photographic exposure or many rapid, sequential exposures (as with a stroboscope). For example, Spedding (1986) used tiny helium bubbles in air as neutral-density particles, which he recorded photographically to visualize the flow. A *streakline* is the pattern formed by a continuous stream of marked fluid released into the flow. Water mixed with dye, released through fine tubing into water flows, can show streaklines in and around organisms (e.g., LaBarbera, 1981), whereas smoke is often used in air to achieve the same effect (Bomphrey et al., 2009). In steady flows, pathlines and streaklines are usually indistinguishable from true streamlines.

smooth, coherent, narrow streak. In turbulent flow, the dye stream quickly breaks up into myriad tiny curls and eddies and rapidly spreads perpendicular to the flow direction. Such dye observations illustrate that the transition from laminar to turbulent flow is usually abrupt, so these two conditions are discrete states; flows are either one of the other, and they spend very little time or space in transition. (A caution: although Reynolds gave a value of 1000 based on radius—usually given nowadays as 2000 based on diameter—as the critical Reynolds number for transition between laminar and turbulent flows in a pipe, this value is only an approximation and can be shifted by almost an order of magnitude in either direction with changes in temperature or wall surface roughness; for external flows—outside of pipes—no single or typical Reynolds number value exists for the laminar—turbulent flow transition.)

Although not quite a molecular level process, the random motion of turbulence is quite small relative to the average motion of a turbulent flow. Such a turbulent flow still gives the general appearance of a mass of fluid all flowing downstream. If you observe the fluid at a near-microscopic level, however, you will see that any given tiny fluid parcel can be moving in almost any direction, with all these tiny, random fluctuations superimposed on the average flow velocity. Whereas the motion of a parcel of fluid in a laminar flow is entirely determined and predictable, the fine-scale motion in a turbulent flow is stochastic.

Finally, velocity gradients near surfaces can be laminar or turbulent as well. We will see in Section 3.3, whether a velocity gradient is laminar or turbulent affects both its thickness and the amount of drag on the surface.

Because the Reynolds number indicates the relative importance of viscous versus pressure drag, it also suggests drag-reduction strategies. At high Reynolds numbers, where pressure drag dominates, the most effective way to reduce drag is to reduce the size of the low-pressure wake, and this can be done by extending and tapering the downstream face of the object (Fig. 3.3). This moves the separation points downstream—and closer together—thus reducing the size of the wake. Shaping to minimize the wake is called "streamlining," and proper streamlining also involves slightly reshaping and extending the front a bit as well to optimize the acceleration of the fluid as it

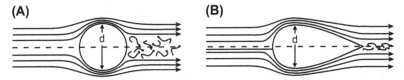

FIGURE 3.3 Wakes that produce pressure drag. (A) Circular cylinder of diameter d. (B) Streamlined strut of maximum thickness d. *From Alexander, D.E., 2016. The biomechanics of solids and fluids: the physics of life. European Journal of Physics 37, 053001,* © *European Physical Society. Reproduced by permission of IOP Publishing. All rights reserved.*

flows over the front and back surfaces. Nevertheless, all streamlined shapes are blunt in front with a long, tapering rear. At high Reynolds numbers, struts with such shapes can have an order of magnitude less drag than a cylindrical strut with a diameter equivalent to the streamlined strut's maximum thickness. Moreover, reversing such a strut, with the tapered end forward and the blunt end aft, can more than double the drag compared to the normal orientation (Hoerner, 1965, p. 6.20). Thus, if flow direction is predictable, or if an organism can reorient to face into a flow, streamlining is a highly effective way to reduce drag. If flow direction is unpredictable or if an organism cannot reorient, a streamlined shape becomes a liability: a streamlined shape broadside to the flow becomes essentially a flat plate with over an order of magnitude more drag than in the "normal" orientation. A cylindrical shape may be the only reasonable option in such flows.

At low Reynolds numbers, in contrast, viscous drag dominates, and the alternate name of this form of drag is skin friction drag for a reason. The more surface area an object has, the more viscous drag it will produce. If viscous drag sufficiently outweighs pressure drag, the added surface area required for streamlining can actually produce increased drag compared to a cylinder or sphere. In these conditions, struts should be cylindrical and bodies should be spherical to minimize drag. This effect is apparent in many organisms; e.g., in aquatic diving beetles (dytiscids), the largest species are conspicuously streamlined, whereas the smallest ones are less streamlined and more spherical.

3.1.5 Drag Reduction in Swimmers and Flyers

For swimming or flying animals, minimizing body drag can be of major benefit. Power is force times velocity, and for a swimmer or flyer, the power needed for locomotion is the drag times the speed of locomotion.

Measuring the drag on a swimming animal can be quite difficult. One particularly useful way to measure the drag on the body of a live, swimming animal is to measure its deceleration between bouts of active swimming. If we know the animal's mass and its deceleration, the drag is the primary decelerating force, so we can calculate this "gliding" drag from Newton's second law. Assuming the animal's propulsors—sea lion flippers, boxfish fins, diving beetle legs—are reasonably well tucked away, this gliding drag is a decent estimate of the drag on the non-thrust-producing part of the body. Body drag can also be measured by towing carcasses or models, measuring sinking speed of weighted carcasses, or placing carcasses or models in flow tanks or wind tunnels. Although subject to additional sources of error (e.g., interference drag, see Tucker, 1990b), these methods can provide useful comparisons for animals with largely inflexible bodies.

In comparing the drag coefficients of a variety of animals of very different shapes across a range of Reynolds numbers, Vogel (1994) used the drag coefficient of a sphere as a blunt or nonstreamlined reference value, and that of a flat plate parallel to the flow as the perfectly streamlined reference value, for any given Re (he converted all drag coefficients to C_{Dw}, drag coefficients based on wetted area). Researchers have used a variety of methods for measuring animal drag. These include gliding or coasting, which give measurements usually termed "whole body drag," as well as towing or flow tank measurements on carcasses or models with propelling appendages removed, where the results are termed "parasite drag" to refer to drag on the non-thrust-producing parts of the body. Other researchers have used calculations based on indirect measures of thrust. Some highly streamlined animals had drag approaching the flat plate drag,[c] such as mackerel, $C_{Dw} = 0.0043$ at Re $= 10^5$ (Webb, 1975); emperor penguins, $C_{Dw} = 0.0027$ at Re $= 1.5 \times 10^6$ (Clark and Bemis, 1979); and California sea lions, $C_{Dw} = 0.0041$ at Re $= 2 \times 10^6$ (Feldkamp, 1987). At the other extreme, frog tadpoles seem to be completely unstreamlined, with C_{Dw} values of 0.1−0.2 at Re between 1000 and 2000 (Dudley et al., 1991), which is essentially the same as a sphere and four times higher than a flat plate in that Re range. Tadpoles tend to live in temporary or very shallow bodies of water, where predatory fish are not normally found, and they eat algae, so have no need to chase food; they apparently face little selection pressure to swim fast or economically.

Fish with flexible bodies seem to have slightly higher gliding drag; for example, in rainbow trout, $C_{Dw} = 0.015$ at Re $= 10^5$ (Webb, 1975), about twice the drag of a flat plate but still eight times less than the drag of a sphere. If, like trout and many other fish, the swimmer uses a substantial portion of its body to produce thrust by undulating, its actual drag while swimming may be quite different from the drag on its body when not swimming. In fact, measuring the drag on a flexible body that is also producing thrust has proven to be nearly intractable, so while the coasting drag is a real cost experienced by a fish, such values may have limited bearing on the drag the fish must overcome when actively swimming. Some recent computer models are beginning to allow researchers to tease apart drag and thrust during undulatory swimming (Chen et al., 2011; Kohannim and Iwasaki, 2014), although application to real fish is so far limited; we will return to undulatory swimming in Section 3.5.

Birds ought to be under significant selection pressure to reduce drag, and most birds certainly look streamlined. Early measurements of body drag on frozen carcasses (with wings removed) sometimes gave drag coefficients closer to spheres than flat plates, e.g., pigeons (Pennycuick, 1968) and vultures (Pennycuick, 1971) with $C_{Dw} \approx 0.06$ over a Reynolds number range of 2×10^5 to 8×10^5. Later drag coefficient measurements on ducks and geese

c. The flat plate drag coefficient is a function of the Reynolds number, and for turbulent flow, $C_{Dw} = 0.072 \, \text{Re}^{-0.2}$, which is used for this comparison, e.g., $C_{Dw} = 0.0052$ at Re $= 5 \times 10^5$; these Reynolds numbers are near the laminar−turbulent transition, and in this range, the laminar flat-plate drag coefficient is a bit lower, $C_{Dw} = 1.33 \, \text{Re}^{-0.5}$ (Vogel, 1994, p. 135).

came out about one-third lower when care was taken to smooth feathers (Pennycuick et al., 1988), but still substantially higher than flat plate drag. By carefully smoothing the feathers, Tucker obtained drag coefficient values lower still for a falcon, $C_{Df} = 0.26$ ($C_{Dw} \approx 0.03$) at Re values ranging from 6×10^4 to 9×10^4 (Tucker, 1990a), better but still slightly closer to a sphere than to a flat plate. Researchers have problems preening the feathers on a frozen carcass to lay as flat and smooth as they do on a living bird (Tucker and Heine, 1990). The parasite drag on bodies of living birds may turn out to be closer to flat plate drag if researchers find a way to measure it with feathers more naturally arranged. Indirect methods, and measurements on models, suggest that problems achieving a lifelike arrangement of feathers may cause overestimates of drag coefficients on feathered carcasses (Tucker, 2000).

Flying insects operate at much lower Reynolds numbers than birds, and fruit flies (*Drosophila* species) may be near the minimum size for effective flight in air. Vogel measured parasite drag on fruit fly bodies and found $C_{Dw} = 0.17$ at Re = 300, about half way between a sphere and a flat plate at that Re (Vogel, 1966, 1994). In contrast, a classic study of locust flight found C_{Dw} values of about the same as a sphere at Re = 8000 (Weis-Fogh, 1956), and indeed, locusts (grasshoppers) show little evidence of streamlining. Apparently factors other than drag reduction played a greater role in the evolution of locust external anatomy.

The results of a study I performed on two species of isopods (small aquatic crustaceans) serve as a cautionary tale about the importance of specifying which reference area is used for C_D calculations, as well as the relative nature of the C_D concept itself. One species was shorter, rounder, wider, and slower swimming, whereas the other species was more elongated, less rounded, and faster swimming. When I measured the average drag on each species and calculated drag coefficients, I found that the short, stout isopod species had a lower C_{Df} (based on frontal area) than the long, narrow one, but a higher C_{Dw} (based on wetted area) (Alexander, 1990) at Reynolds numbers between 3000 and 9000.

3.1.6 Drag as an Asset

At Reynolds numbers below 1.0, drag becomes proportional to speed (instead of speed squared as at higher Reynolds numbers). This direct proportionality leads to some simple but useful relationships, such as Stoke's law for drag on a sphere:

$$D = 6\pi r v \qquad (3.6)$$

where r is the sphere's radius. Similar equations exist for disks and cylinders (Cox, 1970; Vogel, 1994). Only the smallest multicellular animals operate in this regime of thick velocity gradients and dominant viscosity, where surface area matters more than streamlining (although the appendages of many small swimmers fall into this range). Sinking speed, however, is one phenomenon that involves low Reynolds numbers in several macroscopic organisms. For planktonic animals, pollen grains, fungal spores, and wind-dispersed seeds, falling or sinking speed may be biologically important. *Terminal velocity* is the falling speed at which drag, D, exactly balances the force due to gravity. For a sphere, D is given by Stoke's law, and the gravitational force is the weight minus any buoyancy due to displaced fluid. For a sphere at $Re = 0.1$, the terminal velocity, v_T, is

$$v_T = \frac{2r^2 g(\rho - \rho_0)}{9\mu} \qquad (3.7)$$

where g is the gravitational acceleration, ρ is the object's density, and ρ_0 is the fluid's density. If a lower terminal velocity is beneficial, increasing the surface area often evolves: the mass of fine fibers that form the "parachute" of a dandelion seed has exceedingly high surface area, greatly reducing its v_T. Thus, the high-drag parachute of such seeds makes them easy to dislodge in a puff of wind, and their low v_T means that they spend more time being carried by the wind before landing. Tiny spiderlings and very small caterpillars sometimes spin out a single, very long strand of silk that is so long, it has a similar effect, allowing them to ride the wind to new locations (rather inaccurately termed "ballooning").

3.2 FUNDAMENTAL EQUATIONS

3.2.1 Bernoulli's Equation

Bernoulli's equation (or principle) is actually a set of variations on an equation that express the relationship between static pressure, dynamic pressure, and manometric pressure. The derivation is beyond the scope of this book (see Vogel, 1994; Fox and McDonald, 1998); a derivation is sometimes given based on work−energy relationships (Vogel, 1981), but the equation is more fundamentally derived from force and momentum principles (Bertin and Smith, 1979). The most general form of the Bernoulli equation is

$$P + \frac{1}{2}\rho v^2 + \rho g h = \text{constant} \qquad (3.8)$$

where P is static pressure, ρ is fluid density, v is fluid speed, h is height above some datum, and g is the acceleration of gravity. Given its status as a fundamental relationship in fluid mechanics, Bernoulli's equation has some surprisingly restrictive assumptions: no viscosity, steady flow (i.e., no changes with time), and measured along a streamline. The terms in Bernoulli's equation represent energy per unit volume (which has dimensions of pressure) and so the equation can represent a statement of conservation of energy. The static pressure, P, is due to molecular motion and thus represents thermal energy, the dynamic pressure, $^1/_2\rho v^2$, represents kinetic energy, and the manometric pressure, $\rho g h$, represents gravitational potential energy.

A number of variations of the basic equation have practical advantages. For example, if we are interested in comparing two specific locations along a streamline, the constant drops out and we just look at differences between location 1 and location 2:

$$(P_2 - P_1) + \frac{1}{2}\rho(v_2{}^2 - v_1{}^2) + \rho g(h_2 - h_1) = 0. \tag{3.8a}$$

If net changes in height are negligible, $h_2 - h_1 \approx 0$ so the gravitational term can be neglected:

$$\Delta P + \frac{1}{2}\rho\Delta v^2 = 0 \tag{3.8b}$$

where $\Delta P = P_2 - P_1$ and $\Delta v^2 = (v_2^2 - v_1^2)$. In this form, Bernoulli's equation illustrates that pressure varies inversely with the square of speed along a streamline: doubling the speed will produce a four-fold drop in pressure. Finally, if we set $v = 0$ (the fluid is at rest), we get the standard manometric pressure equation:

$$P_2 - P_1 = \rho g(h_2 - h_1). \tag{3.8c}$$

Bernoulli's equation only applies to an incompressible fluid with no viscosity, in other words an "ideal fluid," and then only along streamlines. Given that viscosity—resistance to rate of shear—more or less defines a real fluid, is Bernoulli's equation ever useful? Because a large fluid flow far from walls may experience relatively little shearing, the effects of viscosity will be negligible, and Bernoulli's equation will give useful results. For biomechanics researchers, this usually means fairly large-scale air or water flows well away from solid surfaces, i.e., outside of velocity gradients. The flows well away from the surfaces of a bird's wing or from the body of a swimming porpoise experience little shear, so Bernoulli's equation can be applied with reasonable accuracy. In contrast, flow in tubes, especially tubes of small diameter found in animal circulatory systems and plant vascular transport systems, involves lots of shearing and so is heavily influenced by viscosity; Bernoulli's equation cannot be used to give quantitatively accurate answers. In short, Bernoulli's equation is appropriate for external flows not involving steep velocity gradients, but loses accuracy close to solid surfaces, and may be quite misleading in pipes or vessels of biologically relevant size. Thus, the situations where this fundamental relationship can accurately be applied are surprisingly limited in biomechanics. Vogel (1994, pp. 52–62, 2013, pp. 115–121) gives a more detailed discussion of Bernoulli's equation and its biomechanical limitations.

3.2.2 The Navier–Stokes Equations

As fundamental as Bernoulli's equation is to fluid mechanics, another set of equations is even more basic. The equations of motion for a fluid take the form of a formidably elaborate set of simultaneous partial differential equations called the Navier–Stokes equations. These equations have no analytical solutions except for a couple of situations with very specific boundary conditions, so in the past, most effort went into finding approximations or simplifications. With the advent of powerful digital computers, the Navier—Stokes equations can now be solved numerically for a wide variety of conditions. Nevertheless, such solutions often require supercomputers and even then can take hours or days, with unsteady conditions (as common in biology) greatly increasing the computational difficulties. Analyses based on Navier—Stokes equations are beyond the scope of this book (see Fox and McDonald, 1998).

3.3 VELOCITY GRADIENTS AND BOUNDARY LAYERS

We have seen how the no-slip condition requires a velocity gradient near a solid surface. Regardless of the nature of the solid surface—smooth or rough, hydrophobic or hydrophilic—the fluid velocity will be zero at the surface and increases with distance away from the surface. In this section we will look at the velocity gradient in more detail, especially the factors that affect its thickness.

3.3.1 What Is a Boundary Layer?

A velocity gradient near a solid surface is also called a *boundary layer*. Ludwig Prandtl developed the boundary layer concept to simplify theoretical analyses of fluid movements: in the boundary layer, shear—and hence, viscosity—is significant, but at high Reynolds numbers outside the boundary layer, viscosity can be ignored, which greatly simplifies analyses using the Navier—Stokes (and Bernoulli) equations. So Prandtl defined the boundary layer as the layer from the solid surface out to the distance where the flow speed equals 99% of the free-stream speed. Biologists often have the impression that the boundary layer is a physically distinct fluid feature with some obvious physical demarcation, when in fact it is an arbitrarily defined border in a gradually changing velocity field with no natural discontinuities. In Vogel's words, those researchers "have the fuzzy notion that it's a distinct region, rather than the distinct notion that it is a fuzzy region" (Vogel, 1994, p. 156).

3.3.2 Boundary Layer Thickness

In biomechanics, we are usually most interested in the thickness of a boundary layer. The simplest case is the layer on one side of a flat plate parallel to the flow. The boundary layer gets thicker with distance back from the front or *leading edge* of the plate, and the rate of growth in thickness with distance from the leading edge depends on several factors, including the viscosity, the free-stream fluid speed (v_∞), the fluid density, and whether the boundary layer is laminar or turbulent. We need two versions of the Reynolds number for analyzing boundary layers: the standard one, using the length of the plate from front to back, and the *local Reynolds number*, Re_x, where the reference length is the distance from the leading edge of the plate to a given distance x back from the leading edge. For flat plates with Reynolds numbers ranging from approximately 500 to 5×10^5 and laminar boundary layers, the boundary layer thickness, δ, is given by

$$\delta = 5.0 \sqrt{\frac{x\mu}{\rho v_\infty}} \qquad (3.9a)$$

where x is the distance from the leading edge (Bertin and Smith, 1979, p. 145; Vogel, 1994, p. 158). The definition of the local Reynolds number allows us to substitute in Re_x and get

$$\frac{\delta}{x} = 5.0 \text{Re}_x^{-1/2}. \qquad (3.9b)$$

So the boundary layer thickness forms a parabola with distance from the leading edge, and it will increase in thickness as viscosity increases and decrease in thickness as speed or density increases. Thus, at low Reynolds numbers, boundary layers will be thick, and at high Reynolds numbers, they will be thin.

The equivalent equations for turbulent boundary layer thickness on flat plates at Reynolds numbers $>5 \times 10^5$ is (Fox and McDonald, 1998, p. 438)

$$\delta = 0.38x \sqrt[5]{\frac{\mu}{\rho v x}} \quad \text{or}$$
$$\frac{\delta}{x} = 0.38 \text{Re}_x^{-0.2}. \qquad (3.10)$$

At intermediate Reynolds numbers, a boundary layer can start out laminar and then become turbulent, with the location of the transition being a function of Re_x and surface roughness.

One important difference between a laminar and a turbulent boundary layer is that laminar boundary layers really do act much like infinitesimally thin layers of fluids of different speeds sliding past each other. In contrast, because of the chaotic nature of speeds in a turbulent boundary layer, turbulent boundary layer equations represent average speeds and distances; even deep in a turbulent boundary layer, swirls with local speeds as great as the free-stream speed can exist.

3.3.3 Living in Boundary Layers

For a given flow velocity, boundary layers in air tend to be much thicker than boundary layers in water. Eqs. (3.7) and (3.8) indicate that boundary layer thickness is directly related to viscosity but inversely related to density. Although water is much more viscous than air (30−100 times higher over biologically relevant temperatures), the massive 830-fold difference in density matters much more. Specifically, all else being equal, laminar boundary layers in air will be $\sqrt{15}$ (approximately 4) times thicker than in water (Denny, 1993,

p. 127).[d] To use Denny's example, consider an animal 10 cm downstream from the leading edge of a flat plate in a free-stream flow of 10 cm s^{-1}. If the animal could be displaced by a current of 1 cm s^{-1}, it could be up to 1.3 mm tall in air but only 0.3 mm tall in water (Denny, 1993, p. 128).

For most organisms living in velocity gradients, the actual flow speed at particular heights above the substrate are probably more important than the conventionally computed boundary layer thickness. Such speeds are, however, notoriously difficult to predict computationally, changing as they do from a nearly linear gradient near the surface to asymptotically approaching the free-stream speed farther from the surface. At moderate Reynolds numbers ($500-5 \times 10^5$), the increase in flow speed with height is fairly linear up to approximately 0.4δ, and Denny (1993) gives the following semiempirical relationship for flow speeds in a boundary layer over a flat plate:

$$\frac{v_z}{v_\infty} = 0.32z\sqrt{\frac{\rho v_\infty}{x\mu}}, \quad z < 0.4\delta \tag{3.11}$$

where v_z is the flow speed at height z above the surface, at a distance x from the leading edge.

Keep in mind that these equations are all for rather idealized conditions: few organisms, or the surfaces on which they sit, consist of smooth, flat plates with straight leading edges. In nature, boundary layers may be turbulent at lower-than-expected Reynolds numbers (Grace and Wilson, 1976), and surface roughness, irregularities, and protrusions may all affect the boundary layer in unexpected ways (Grace, 1977).

Laminar boundary layers can have great consequences for tiny organisms. If an organism such as a small insect larva or fungal spore is small enough to stay well within the boundary layer, wind or water currents are not likely to dislodge it from the surface. Similarly, pollen grains, tiny mites, or insect larvae that depend on wind for dispersal may need special adaptations to get far enough through the boundary layer to reach useful wind speeds. For example, a 0.1-mm-tall scale insect is too deeply embedded in the boundary layer to be carried away from the surface of a leaf if it is more than a few micrometers from the leading edge of a leaf. By rearing up on its hind legs, it can get enough of its body area into the upper boundary layer where wind speeds are high enough to carry the insect off the leaf, even several millimeters from the leaf's leading edge (Washburn and Washburn, 1984). On the other hand, if a small organism needs access to flowing fluid, for gas exchange surfaces or filter-feeding structures, then the organism will need to be tall

---

d. Technically, turbulent boundary layers will be $15^{-0.2} \approx 1.7$ times thicker in air than water, but due to the great variation in local flow speeds in a turbulent boundary layer, they do not offer small organisms the same opportunity to "hide" or danger of getting "trapped" in the layer as for laminar boundary layers.

enough to reach through most of the boundary layer or select regions where the boundary layer is appropriately thin. Barnacles are suspension feeders with feathery appendages they extend to catch food particles. Some species of barnacles are only found on whales, and they normally settle on the whale's head, where boundary layers will be relatively thin as the whale swims (Blokhin, 1984).

Some open ocean seabirds (e.g., albatrosses) use boundary layers on a much larger scale. When winds blow over ocean surfaces for hundreds or thousands of kilometers, detectable velocity gradients several meters thick can build up. Some seabirds exploit these boundary layers by using a specialized form of gliding called dynamic soaring. They descend in a glide and accelerate downwind, then turn upwind and climb sharply, maintaining airspeed as they climb up through the boundary layer. The boundary layer allows them to trade momentum gained going downwind for potential energy when they turn upwind. The details of dynamic soaring are beyond the scope of this book (see Alexander, 2002 for details); for our purposes, we simply need to know that the behavior will only work in a boundary layer several meters thick.

Prandtl had sound reasons to propose his 99% criterion for boundary layer thickness, but it is nevertheless an arbitrary choice that is largely an artifact of our base 10 number system. Vogel (1994, 2013) has argued that in many biological situations involving laminar velocity gradients, a boundary layer thickness of 90% of free-stream speed is more biologically meaningful: if an animal can get its gills or filtering appendages into water at 90% of v_∞, the extra stretch to get them into water at 99% may not be worth the effort. This "90%" criterion yields a layer with thickness of only 70% of the thickness of a conventional "99%" boundary layer. Replacing the constant of 5.0 with 3.5 in Eqs. (3.9a) and (3.9b) gives δ for a boundary layer with a thickness based on 90% of free-stream speed.

Although we have so far focused on biological consequences of laminar boundary layers, laminar boundary layers can be detrimental compared with turbulent boundary layers in situations where some quantity gets depleted near a surface, e.g., oxygen near the surface of a gill. In a laminar boundary layer, if oxygen is depleted near the surface, a stagnant layer can build up because flow cannot cross streamlines, so fresh, oxygen-laden water cannot move toward the surface. Turbulent boundary layers, in contrast, have considerable flow perpendicular to the mean flow direction, even deep in the layer. Turbulent boundary layers thus keep the flow stirred up near the surface and inhibit stagnation. Size works against turbulence on gills, however; only the largest fish have flow through their gills at high enough Reynolds numbers to permit turbulence. Fortunately, at the small scales of low-Reynolds number gills, diffusion can be rapid enough to cross the direction of boundary layer flow (see discussion of Peclet number in Vogel, 2013, p. 159).

3.4 WINGS AND LIFT

When an object moves through a fluid, if it has the appropriate shape and orientation, the flow field can produce a force at right angles to the direction of motion. This force is called *lift*. Most people probably associate lift with wings moving through air, but many aquatic animals have appendages that produce lift underwater, such as tuna tails and sea lion flippers (such underwater "wings" are usually called hydrofoils). Puffins and dippers even use their wings for both flying in air and swimming under water! Note that lift is defined as being perpendicular to the direction of wing movement. When a wing moves horizontally, lift is vertically upward, but if a wing moves at any angle other than horizontally, lift will *not* be directly upward. This distinction may not matter much for airplanes, whose wings spend most of their time moving more or less horizontally, but it matters a great deal for biological wings, which may spend little or no time moving horizontally, as we will see.

3.4.1 The Lift Mechanism and the Bound Vortex

A well-designed wing is a remarkable device. For the cost of pushing on it to overcome a unit of drag force, it rewards us with many units of lift force. Drag is parallel to the direction of motion and in the opposite direction, so lift and drag are always at right angles to each other.

Wings produce lift by causing the air to move faster over the top and slower under the bottom. Bernoulli's equation says that pressure will be inversely proportional to v^2, so a relatively small difference in flow speed between the top and bottom can produce a substantial difference in pressure. This pressure distribution, integrated over the whole surface of the wing, results in a net upward force on the top of the wing (Fig. 3.4). Note that the wing cross section or *airfoil* shown in Fig. 3.4 is more convex on top than on the bottom.

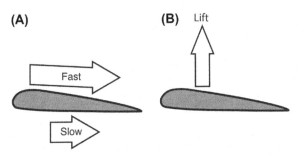

FIGURE 3.4 Airflow over a wing (looking at the cross section from front to back). (A) The differences in airspeed over the top and under the bottom produce differences in air pressure. (B) The differences in pressure sum to yield a net upward force, *lift*. *Redrawn from Alexander, D.E., 2002. Nature's Flyers: Birds, Insects, and the Biomechanics of Flight. Johns Hopkins University Press, Baltimore, Maryland, 358 pp. Artist: Sara Taliaferro.*

Elementary science texts often say that this convexity is what causes lift: if two little packets of air reach the leading edge of the wing at the same time, the upper one going over the top has a greater distance to travel than the lower one, so the upper one must travel faster if they are to reach the back (trailing edge) of the wing at the same time. There are at least two problems with this statement. First, wings can produce lift when upside down (air shows would be much less exciting if this were not true). Second, wind tunnel tests using puffs of smoke on wings producing high lift clearly show that the air over the top arrives at the trailing edge well *before* the air underneath. In fact, if this were not true and the top and bottom air packets did meet at the trailing edge, the speed difference would be nowhere near large enough to produce the pressure difference typically seen on effective wings.

Some specialized terminology is associated with wings. We have already encountered the *airfoil*, the wing's cross section from front to back ("airfoil" originally meant the entire wing, but nowadays in engineering usage it applies only to the wing's cross section). A straight line connecting the front of the airfoil's leading edge to the back of the trailing edge is the *chord*, *c*, which can refer either to the direction of this line or its length (Fig. 3.5). The upward convexity of an airfoil is its *camber*. The camber is measured by drawing the mean line (a curve through the midpoints between the upper and lower surfaces) and measuring the maximum height of the mean line, *h*, above the chord; camber is usually expressed as a percentage of the chord:

$$\text{camber} = \frac{h}{c} \times 100.$$

(All animal wings and most airplane wings are cambered, which increases lift production as long as the wing is right side up; we will see in Section 3.5 that underwater wings or hydrofoils usually produce lift alternately on the top or bottom and thus have either no camber or reversible camber.) Finally, wings

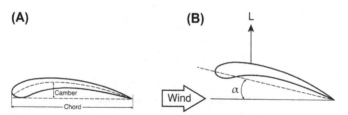

FIGURE 3.5 Airfoil terminology: a wing's cross section from front to back is its airfoil shape. (A) The chord is the length of an airfoil from front to back; it may also refer to a line connecting the frontmost point on the airfoil with the rearmost part of the trailing edge. Camber is the degree of upward convexity, expressed as the maximum height of the airfoil's mean line as a percent of the chord. (B) The angle of attack, α, is the angle between the airfoil's chord and the direction of motion (or relative wind). As the angle of attack increases, lift (*L*) increases. *Artist: Sara Taliaferro.*

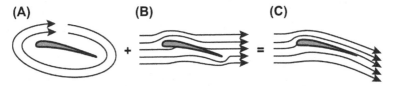

FIGURE 3.6 Mathematically combining a bound vortex, (A), with a symmetrical ideal fluid flow, (B), gives the actual flow pattern around an airfoil, (C). Note that (A) is an essential component of (C) and can never occur alone; (C) is the pattern around any wing producing lift, whereas the flow pattern around a wing at its zero-lift angle of attack will be much like that of any streamlined shape, such as that of Fig. 3.3C. *Redrawn from Alexander, D.E., 2002. Nature's Flyers: Birds, Insects, and the Biomechanics of Flight. Johns Hopkins University Press, Baltimore, Maryland, 358 pp. Artist: Sara Taliaferro.*

usually operate with their leading edges tilted up slightly relative to the direction of motion, thus angling the chord up. The angle between the chord line and the direction of motion is the wing's *angle of attack* (Fig. 3.5).[e]

The key to a wing's lift production is its airfoil's sharp trailing edge. When a wing with a sharp trailing edge is cambered or at a positive angle of attack, the only flow pattern that allows air to flow smoothly off the top and bottom of the trailing edge with no discontinuities is one where the air over the top accelerates to a much higher speed than the air under the bottom. Such a flow pattern can be mathematically partitioned into a symmetrical flow, equivalent over the top and bottom, and a *bound vortex* (Fig. 3.6). (A vortex is a whirling flow pattern such as a tornado or a whirlpool). When summed vectorially, these two flow patterns give the actual flow pattern around the wing. Although the bound vortex is a real phenomenon, it can never exist on a wing by itself, and no little packet of fluid actually circumnavigates the entire wing. It can only exist as a component of the flow along with the symmetrical component, or in Vogel's words, as a sort of "net vortex" (Vogel, 1994, p. 231).

The bound vortex is important in several ways. The strength of the bound vortex, its *circulation*, Γ, is directly proportional to the amount of lift produced. Also, fluid theory does not allow vortices to begin or end in the fluid, so when a wing starts from rest and begins producing lift, it sheds a starting vortex (turning in the opposite direction from the bound vortex), which is connected to the bound vortex by a pair of *tip vortices* (Fig. 3.7). The starting vortex will eventually be dissipated by viscosity, but tip or *trailing* vortices continue to stream off the wingtips as long as the wing produces lift. When a wing stops producing lift, it sheds the bound vortex as a *stopping* vortex; large or fast flyers may maintain the bound vortex throughout the stoke cycle, but

---

e. Biologists often confuse the "angle of attack" with the "angle of incidence." The incidence angle is the angle of a wing's chord with respect to some reference line on an airplane's fuselage; this angle is thus defined by the airplane's structure and not its movement through the air. Because of their mobile wings, animals do not normally have a fixed incidence angle.

(A)   **(B)**

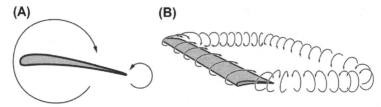

FIGURE 3.7 (A) Cross section of bound and starting vortex on an airfoil at the instant it begins to produce lift. (B) Schematic view of the three-dimensional vortex system of a wing. *Redrawn from Alexander, D.E., 2002. Nature's Flyers: Birds, Insects, and the Biomechanics of Flight. Johns Hopkins University Press, Baltimore, Maryland, 358 pp. Artist: Sara Taliaferro.*

smaller or slower flyers may shed stopping vortices at the end of the downstroke, or in some cases at the end of the downstroke and upstroke. Moreover, when animals flap their wings, we will see that they sometimes take advantage of modifications to these vortices to enhance lift production.

Note that on the outside of a tip vortex, the air is moving upward (Fig. 3.7). Birds that fly in formation appear to take advantage of this: if their spacing is just right, a bird alongside of and slightly above and behind another bird can get a little upward boost from the tip vortex of the bird in front (Hainsworth, 1989; Speakman and Banks, 1998; Maeng et al., 2013). The "V" formation used by migrating geese appears to put individuals in position to take at least some advantage of this added boost.

Just as with drag, a nondimensional index of lift is useful for comparisons. The *lift coefficient*, C_L, is given by

$$C_L = \frac{L}{\frac{1}{2}\rho v^2 S} \tag{3.12}$$

where L is the magnitude of the lift. The reference area in this case is always the *planform area*, i.e., the projected area of the wing looking down on it from above. By convention, the planform area includes the area that connects the right and left wing roots through the animal's body or the airplane's fuselage. (As with the drag coefficient equation, rearranging Eq. (3.12) to put L on one side of the equation does not define lift, despite it often being called the "lift equation.") Lift coefficients a bit above or below 1.0 are typical for animal wings, and lift coefficients ≥ 2.0 indicate something other than conventional aerodynamics based on bound vortices.

3.4.2 Modifying Lift

Figs. 3.6 and 3.7 illustrate another important aspect of lift production by wings. Note that the air leaves the trailing edge of the airfoil in Fig. 3.6 angled slightly downward. This downward deflection is the *downwash*, and another way of thinking of lift production is as an example of Newton's third law: "for every action there is an equal and opposite reaction." Deflecting the air downward is the action, upward force on the wing is the reaction. The vortex system on the wing also implies the downwash, with the vortex turning down on the trailing edge and on the inside of the tip vortices.

What if the shape or orientation of the wing were changed so that the trailing edge tilted down a bit more? That would increase the downwash, which is another way of saying that it would increase the circulation (strength of the bound vortex) and hence, lift. One way to do this is to increase the camber; the more "humped" the shape of the airfoil, the more lift it produces at zero angle of attack. A little camber is useful because it allows a wing to produce more lift at zero or low angles of attack, but 15% would be considered very high camber. Above that, increased drag and increased tendency for flow separation begin to outweigh any lift enhancement.

The other way to tilt the trailing edge down is by increasing the angle of attack. Within certain limits, the lift will be approximately proportional to the angle of attack. One limit to increasing the angle of attack to produce more lift is a phenomenon called *stall*. Above some critical angle of attack, separation occurs, where the airflow over the top of the wing peels away and flows straight back rather than following the wing upper surface (Fig. 3.8). Circulation and downwash largely disappear, so most lift is lost, and a large, turbulent wake forms, greatly increasing drag. In a nutshell, a stalled wing suddenly has very little lift and a lot more drag. Generally, stalls are to be avoided, although a stall just at touchdown during landing can be desirable.

In addition to camber and angle of attack, the wing's area and the wing's speed also directly affect the amount of lift it produces. The effect of wing area is intuitively reasonable, because more wing should produce more lift, and in fact, the amount of lift produced by a wing is directly proportional to the wing's planform area. Conventional airplanes make relatively little use of wing area changes to adjust lift (although research into "morphing wings" may someday change this (Li and Ang, 2016; Previtali et al., 2016)), whereas flying animals may make large lift adjustments by changing the area of their wings. Gliding birds, for example, are well known to extend or flex their wings to adjust lift (Tucker and Parrott, 1970). (Large airliners have multielement

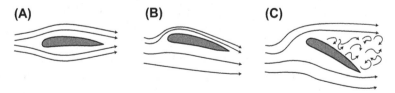

(A)   **(B)**   **(C)**

FIGURE 3.8 Progressively increasing the angle of attack from (A) to (B) to (C) eventually results in separation and stall (C). *From Alexander, D.E., 2002. Nature's Flyers: Birds, Insects, and the Biomechanics of Flight. Johns Hopkins University Press, Baltimore, Maryland, 358 pp. Artist: Sara Taliaferro.*

Fowler flaps that can increase wing area by 10% or 20% (Bertin and Smith, 1979, p. 379), but the flaps of smaller airplanes, if present at all, affect camber much more than wing area; in any case, even a 20% change is meager by birds' standards).

Speed's effect is a bit different; lift is proportional to the square of speed, v^2. Small changes in speed can thus produce relatively large changes in lift production. Given that lift is the net result of pressure differences on the top and bottom of the wing and that Bernoulli's equation relates pressure to v^2, the relationship between lift and speed again seems logical.

All these factors affecting lift interact, and all flying animals can and do use them simultaneously to adjust lift. If an animal attempts to maintain level flight after speeding up, it will have to use some combination of decreased camber, angle of attack, and wing area to avoid climbing. Animals generally make much more use of changes to wing geometry for lift adjustments than airplanes do.

3.4.3 Induced Drag Causes and Consequences

The process that generates lift also generates a new form of drag, the *induced drag*, D_i. The induced drag increases with increasing angle of attack (although not in a simple or linear manner), meaning that induced drag increases as lift increases. The induced drag thus represents the cost of lift production. Although lift is obviously necessary for flight, a wing's effectiveness is actually better characterized by its lift-to-drag ratio, L/D, than by the amount of lift alone. A wing might produce a huge amount of lift, but if it also produced a huge amount of drag, it would not be a very useful wing. An ordinary, garden variety bird or airplane wing might have an L/D of 10, meaning that it only needs to overcome, say, 10 N of drag to get 100 N of lift. The work to fly goes into thrust production, which is needed to overcome drag, so the L/D is the main figure of merit for a wing. The L/D is not constant for a given wing, however, but changes with the angle of attack. Researchers thus normally use the maximum lift-to-drag ratio, L/D_{max}, to characterize a given wing. I leave it as an exercise for the reader to show that L/D is identical to C_L/C_D.

Gustav Eiffel, designer of a famous Parisian landmark, developed a clever way to display the relationship between a wing's lift and drag, using a so-called polar plot (Fig. 3.9).[f] Such a graph plots C_D, on the horizontal axis, versus C_L, on the vertical axis, and shows values of the angle of attack parametrically on the curve. A polar plot illustrates several important wing properties. The far right end of the curve shows the critical angle of attack, i.e., the angle of attack at stall. A line from the origin tangent to the curve has a slope with the

---

f. Despite the rather unfortunate name, a lift-drag polar plot is drawn in Cartesian, not polar, coordinates.

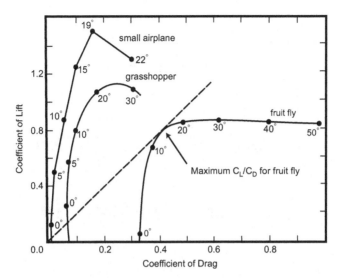

FIGURE 3.9 Polar diagram for wings of a small airplane, a grasshopper, and a fruit fly. Angles of attack are shown on each curve. The upper right-hand end of each curve represents the point where stall occurs. A straight line tangent to the curve that passes through the origin touches the curve at the angle of attack where the maximum C_L/C_D occurs (e.g., dashed line, 18 degrees for fruit fly). Note that as size (and hence Reynolds number) decreases, the maximum C_L decreases, but stall is also delayed to higher angles of attack. *Redrawn by Sara Taliaferro from Alexander, D.E., 2002. Nature's Flyers: Birds, Insects, and the Biomechanics of Flight. Johns Hopkins University Press, Baltimore, Maryland, 358 pp.; based on data in Vogel, S., 1967. Flight in* Drosophila. *III. Aerodynamic characteristics of fly wings and wing models. Journal of Experimental Biology 46, 431–443.*

maximum *L/D*, and where this line is tangent to the curve gives the angle of attack at maximum *L/D*. The overall maximum lift and minimum drag are shown by the horizontal and vertical tangents to the curve, respectively. Note that the maximum lift almost never occurs at the angle of attack of the maximum *L/D*; the maximum *L/D* normally occurs at a much lower angle of attack than the maximum lift. Thus, flying at the angle of attack of maximum lift is much more expensive than flying at the angle of maximum *L/D*.

One of the key factors influencing the lift-to-drag ratio is a geometric property called the *aspect ratio, AR*. The aspect ratio quantifies whether a wing is long and narrow—like a gull's high aspect ratio wing—or short and broad, like a sparrow's low aspect ratio wing (Fig. 3.10). Conceptually, the aspect ratio is the ratio of the wingspan (wing length from tip to tip), *b*, to the wing chord, *c*, and for a rectangular wing, *b/c* is exactly the aspect ratio (Fig. 3.10). No flying animal has a perfectly rectangular wing, so for nonrectangular wings we use

$$AR = \frac{b^2}{S} \tag{3.13}$$

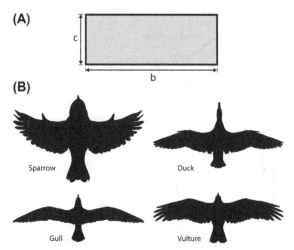

FIGURE 3.10 Aspect ratios. (A) Top view (planform) of a rectangular wing, where the aspect ratio is the span, *b*, divided by the chord, *c*. (B) Planform silhouettes of a variety of birds, drawn so they all have the same wingspan (and are thus not to scale), to illustrate variation in aspect ratio. *Artist: Sara Taliaferro.*

(where S is planform area), which essentially gives the ratio of the wingspan to the mean chord.

The aspect ratio affects the lift-to-drag ratio because the induced drag is related to processes that take place at the wing's tip, and the more "tip" a wing has for a given area, the greater the induced drag. So wings of low aspect ratio—lots of tip for their size—have higher induced drag than wings of high aspect ratio. A high induced drag in turn reduces the L/D, so low aspect ratio wings tend to have low L/D, and high aspect ratio wings tend to have high L/D, all else being equal. For example, a sparrow with an aspect ratio of approximately 5 might have an L/D of 4, whereas a swallow of similar body weight but with wings with an aspect ratio of 11 would have an L/D of 10 (Tennekes, 1996). Any wing is a compromise, and sparrows have evolved to be highly maneuverable in cluttered environments, such as trees and underbrush, where long, narrow wings would be impractical. Swallows, in contrast, are adapted for fast, continuous flight in open habitats, where aerodynamic efficiency is vital.

Some gliding birds—vultures, hawks—glide with the tip or "primary" feathers distinctly separated (Fig. 3.10, vulture). These separated primaries, also called slotted tips, give the wing a slightly higher L/D than might be expected based on the wing's aspect ratio (Tucker, 1993, 1995). Although the aerodynamics are more complicated than this, a simple way to think about separated primaries is to consider each primary feather as its own little wing, so that what would otherwise be a broad, blunt tip is divided up into several smaller, higher aspect ratio wings. Experiments clearly demonstrate the effectiveness of separated primaries, but researchers do not entirely agree on exactly how they work.

Flying animals are masters of changing their wing geometry to adjust aerodynamic forces. Birds and bats can flex their wings to reduce or increase their area, and bats and especially insects can increase or decrease the camber of their wings. (We will see in the next section that some insects can even reverse the camber of their wings during flapping.) Airplanes make, at most, modest use of such geometric changes (e.g., changing camber and sometimes area slightly using wing flaps), but work on "morphing wings" is an active area of engineering research that is seeking mechanisms that will allow engineered wings to take advantage of some of the shape changes used by flying animals (e.g., Li and Ang, 2016; Previtali et al., 2016).

3.4.4 Gliding and Flapping

Contrary to popular belief, birds do not flap their wings to produce lift, at least not directly. Flying animals, whether mosquitoes or eagles, flap their wings to produce *thrust*. Wings produce lift simply by moving through the air: airplanes do not flap their wings, and those wings produce lift quite nicely. Similarly, an animal can stop flapping with its wings extended and glide, and its wings produce plenty of lift while gliding.[g] Gliding is actually gravity-powered flight. By following a flight path angled slightly downward, gravity essentially pulls the glider along. As long as the vector sum of lift and drag, the so-called *resultant force*, *R*, exactly balances the glider's weight (Fig. 3.11), the glider can glide until it runs out of altitude (see Box 3.4). A glider cannot go upward with respect to the air, i.e., its path through the air must always be downward. If, however, the air itself is rising, a glider can be carried upward if the air is rising faster than the glider is descending. This is called *soaring*, and soaring is how vultures and hawks are able to stay aloft, and even ascend, without flapping their wings. As long as they can find air rising faster than their sinking speed, they can fly with very little effort. Sources of rising air include thermals (columns or bubbles of warm air) and "orographic lift," which is wind blowing up the slope of a hill or mountain, also called "ridge lift" (Alexander, 2002). Soaring is a specialized behavior; not all flying animals can soar, and even for those that can, rising air is not always available.

As mentioned above, flying animals flap their wings to produce thrust, a forward force. Thrust is what pushes a flyer forward through the air, and wings moving forward through the air will automatically produce lift. Fixed-wing airplanes have separate engines to produce thrust, but flying animals must

---

g. When biologists compare gliding with what they call "true flight" (i.e., flapping or powered flight), they show lack of understanding of wing operation. The wings of a gliding bird are doing exactly the same thing as the wings of a flying airliner, and no one would deny that the airliner is in "true" flight.

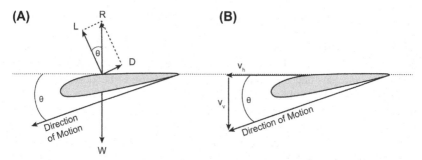

FIGURE 3.11 Geometry of gliding. (A) The vector sum of the lift (*L*) and drag (*D*) is the resultant force, *R*. The resultant force exactly balances weight, *W*, when the glide angle of the wing, θ, is the same as the angle between *L* and *R*. (B) The glide angle sets the glider ratio, which is the ratio between the horizontal speed, v_h, and the vertical or sinking speed, v_v. *Modified from Alexander, D.E., 2002. Nature's Flyers: Birds, Insects, and the Biomechanics of Flight. Johns Hopkins University Press, Baltimore, Maryland, 358 pp. Artist: Sara Taliaferro.*

BOX 3.4 Gliding for a Living

When a glider is flying in equilibrium, it will glide with a steady (constant) speed and a constant *glide ratio*. The glide ratio is the ratio of the distance a glider moves forward, d_h, to the distance it descends, d_v, in a given time (Fig. 3.11); since distance per time is speed, this turns out to be the same as the ratio of the horizontal speed, v_h, to the vertical or sinking speed, v_v. In an equilibrium glide, the geometry requires the glide ratio to be equal to the lift-to-drag ratio (see Alexander, 2002 for details). Thus, the *L/D* of a glider sets the glide ratio, which can also be expressed as the *glide angle*, θ (Fig. 3.11), where $\theta = \cot^{-1}$ (glide ratio).

Most flapping flyers can stop flapping and glide, but a surprising variety of animals (and some plants!) is "pure gliders," that is, they can glide but cannot flap. Familiar examples include flying squirrels, flying fish, and maple and poplar seeds. More exotic gliders include gliding lizards, gliding tree frogs, and even gliding snakes (Alexander, 2002, 2015). Because of constraints on their wing surfaces, none of these are stellar gliders; flying squirrels have lift to drag ratios of only 2 or 3 (Thorington and Heaney, 1981), well below that of even the least efficient bird wings. Nevertheless, under the right conditions, many pure gliders can achieve much flatter glides than their lift-to-drag ratios would suggest.

In fact, flying squirrels and flying lizards often avoid equilibrium gliding. By starting out fast, with a quick drop to build up airspeed, they can decelerate while constantly increasing their angle of attack and plan to stall just as they reach their intended landing spot. (Their low aspect ratio wings can reach unusually high angles of attack before stalling (Alexander, 2002).) Pulling off such a precise maneuver requires a sophisticated ability to judge distances and fine aerodynamic control, but most such purely gliding animals seem to perform it routinely.

produce both lift and thrust using only their wings. Functionally, a crow's or dragonfly's wings have more in common with a helicopter rotor than with airplane wings. Like the flapping wings of a bird, a helicopter's rotor must also produce both lift and thrust.

Given the huge size range of flapping flyers, from fruit flies to condors, the basic flapping pattern is surprisingly similar. In normal forward flight, the downstroke generates most of the useful forces. During the downstroke, the wings move down and forward with the chord nearly horizontal or with the leading edge tilted slightly down. The wing moves downward at a slightly steeper angle than the wing's chord, so it maintains a positive angle of attack. The lift is perpendicular to the wing's motion so it is tilted forward, and this forward tilt is key. The vertical component of lift supports the animal's weight, but the forward component represents thrust (Fig. 3.12). For steady, level flight, the mean thrust must equal the total drag of the flyer. In forward flight, most or all of the useful forces are generated during the downstroke.

Different flyers do different things during the upstroke. Large, fast-flying birds, such as geese, cranes, or gulls, generally use a low-force upstroke that mainly serves as a recovery stroke to get the wing in position to perform the next downstroke. Sometimes called a "passive" upstroke, the bird orients its wing to minimize aerodynamic forces on it—it may produce a small amount of lift or it may just produce the minimum possible drag (Fig. 3.13A). If a bird uses a passive upstroke, the mean lift and thrust of the downstroke alone must be equal to the mean weight and drag of the entire stroke.

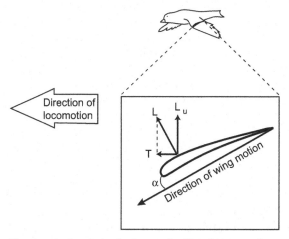

FIGURE 3.12 Thrust production during the downstroke. The wing moves down and forward at a positive angle of attack, producing lift, *L*, at right angles to the wing movement. The upward component of lift, L_u, supports the animal's weight, while the forward component represents the thrust, *T*, that overcomes the animal's drag. *Redrawn from Alexander, D.E., 2016. The biomechanics of solids and fluids: the physics of life. European Journal of Physics 37, 053001, © European Physical Society. Reproduced by permission of IOP Publishing. All rights reserved.*

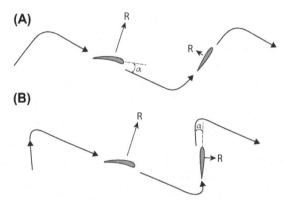

FIGURE 3.13 Wingbeat patterns showing path of the wingtip through the air and the aerodynamic force on the wing, using different upstrokes. (A) Inactive upstroke, used by large or fast flying animals: wing at a very low angle of attack during upstroke to minimize forces. (B) Same downstroke but with active upstroke, used by small or slow flying animals: wing operates at a negative angle of attack with reduced or reversed camber during upstroke, and the force on the wing is largely in the thrust direction. *R*, resultant force. *Modified from Alexander, D.E., 2002. Nature's Flyers: Birds, Insects, and the Biomechanics of Flight. Johns Hopkins University Press, Baltimore, Maryland, 358 pp. Artist: Sara Taliaferro.*

Very small flyers, particularly insects and probably hummingbirds and small bats, use an active upstroke. To use an active upstroke, the flyer needs to be able to twist its wings so the chord is approximately vertical, and it must be able to move its wings up and back faster than it is flying forward through the air (Fig. 3.13B). If it is capable of these movements, the anatomical lower surface functions as the top of the wing—so the ability to reduce or reverse the wing's camber is also beneficial—and lift is produced almost directly forward (Fig. 3.13B). The useful force during such an upstroke is thus almost entirely thrust. Because of the need to move the wing backward so fast, active upstrokes are always of much shorter duration than the same animal's downstroke. So while an active upstroke does generate useful thrust, both the magnitude of the thrust and its duration is substantially lower during the upstroke than the downstroke. And of course, little or no upward force is produced during such an active upstroke.

Although I have described active and passive upstrokes as if they are two separate modes, they may actually be extremes on a continuum. Many medium-sized flying animals probably use some intermediate form of upstroke that generates a bit of extra lift or a bit of extra thrust, or both.

My description of flapping wings has so far treated them as if they work just like airplane wings, and in general, they do. In detail, however, animal wings sometimes show significant differences (Ellington, 1984c; Van den Berg and Ellington, 1997a). One difference is the *leading edge vortex*, first clearly demonstrated by Charles Ellington and colleagues on a robotic flapping model of a moth's wing (Ellington et al., 1996; Van den Berg and Ellington, 1997b).[h] Rather than a bound vortex surrounding the wing, a vortex forms that sits just above and behind the wing's leading edge, with air in the core spiraling out from near the wing base to the tip. Such a vortex has more or less the same effect as a bound vortex. Since being described by Ellington and coworkers, researchers have observed leading edge vortices on other insect wings (Sane, 2003) and on the wings of swifts (Videler et al., 2004). Leading edge vortices seem to be linked with wings having leading edges that are both sharp and angled posteriorly ("swept back"). Although research on the significance of leading edge vortices in animal flight is ongoing, they seem to allow wings to operate at higher angles of attack without stalling and to form more quickly than bound vortices. The latter would be a significant benefit to an animal like a house fly that flaps its wings over 100 times per second and may shed the vortices from its wings once or twice per wingbeat.

3.4.5 Hydrofoils

As mentioned earlier, wings can work just as well under water as in air, although such submerged winglike structures are usually called *hydrofoils*.[i] Many swimming animals use appendages for locomotion that operate essentially the same way as a flapping wing. The front flippers of a sea turtle illustrate the process nicely. A turtle flaps its flippers up and down, with a positive angle of attack on the downstroke and a negative angle of attack on the upstroke. The upstroke and downstroke are basically mirror images of each other (Fig. 3.14) with a forward component in both halves of the stroke and vertical components that cancel each other on each stroke half. Such hydrofoil appendages are often symmetrical and generate lift entirely using angle of attack, as shown in Fig. 3.14. Some may be able to reverse the camber by twisting at the base and allowing some passive deformation from hydrodynamic loads—as done by insects to improve their active upstroke—but my impression is that hydrodynamic loads are so great that most underwater flappers use rigid, symmetrical profiles to provide the needed strength to resist those loads. Sea lions and penguins use a very similar flapping stroke for swimming propulsion.

Cetaceans—dolphins, porpoises, and whales—swim using their tail fins or "flukes" as hydrofoils. Rather than flapping each fluke in an angular motion around a hinge at the root, a porpoise moves its entire tail up and down as a unit. Nevertheless, the pattern of force production is largely the same, with lift

---

h. Leading edge vortices had been known to engineers from observations on delta-winged aircraft since the mid-20th century, but had not been associated with flying animals until observed in Ellington's lab.

i. Confusingly, both a watercraft that lifts most of its hull out of the water using submerged wings, and such a craft's wings themselves, are referred to as "hydrofoils." For our purposes, "hydrofoil" will be synonymous with underwater wing.

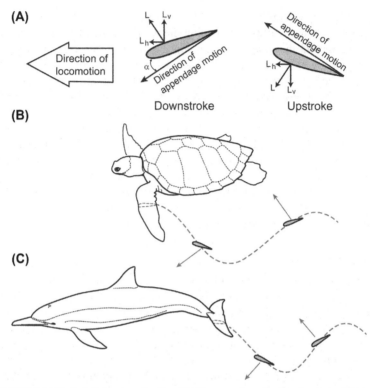

FIGURE 3.14 Hydrofoil thrust production. (A) Cross section of a flapping hydrofoil showing forces during upstroke and downstroke. Because it produces lift (L) on the upper surface during the downstroke and the lower surface on the upstroke, the hydrofoil is uncambered; note positive angle of attack (α) during downstroke and negative angle of attack during upstroke. Vertical components of lift (L_v) of downstroke and upstroke cancel; horizontal components of lift (L_h) act as thrust. (B) Swimming movements of sea turtle's front flippers. Dashed line shows path of one flipper through the water; outline of cross section shows the orientation of hydrofoil at an instant during the downstroke and an instant during the upstroke, with arrows showing the net lift at that instant. (C) Swimming movements of a dolphin's tail, forces and movements as shown in part (B). *Artist: Sara Taliaferro.*

angled forward on both the upstroke and downstroke (Fig. 3.14C). Take the porpoise tail beat pattern and turn it on its side so the tail moves horizontally, and it becomes the tail beat mechanism for tunas, swordfish, large sharks, and other large, fast-swimming fish. Finally, we will see in Section 3.5 that even some swimmers without obvious winglike appendages can use lift production for locomotion.

3.4.6 Wings and Size

The effectiveness of a wing, embodied by the lift-to-drag ratio, decreases as the Reynolds number decreases. At high Reynolds numbers, skin friction is low and velocity gradients are thin, and an airfoil's streamlined shape minimizes pressure drag. At low Reynolds numbers, viscous drag increases, thus increasing total drag. Moreover, at low Reynolds numbers, boundary layers thicken, meaning that the shape of an airfoil has less influence on the flow, and bound vortices do not form as easily, so circulation is less. Thus, even for geometrically similar wings, as the Reynolds number drops below approximately 10^3, simultaneous increase in drag and decrease in lift means that wings' lift-to-drag ratios drop substantially. Researchers tested a model wing with a standard technical airfoil at Reynolds numbers down to 1.0, and their airfoil model had lift-to-drag ratios of approximately 1.4 at Re = 100, 0.43 at Re = 10, and a dismal 0.18 at Re = 1.0 (Thom and Swart, 1940).[j] These measurements correspond reasonably well with lift-to-drag ratios of small biological wings, including those for locusts (large grasshoppers) at approximately 8 (at Re = 4000) (Jensen, 1956) and bumblebees at 2.5 (at Re = 1200) (Dudley and Ellington, 1990). Fruit flies, with a miserable *L/D* of 1.8 (Re = 200) (Vogel, 1967), may be approaching the lower limit for effective wing-powered locomotion. Whereas a large hawk might have a maximum lift-to-drag ratio of 9 or 10 at an angle of attack of 6 degrees (and a stall angle of 15 degrees), the fruit fly's wing had its maximum lift-to-drag at 15 degrees and showed no sign of stalling below 50 degrees. This compares with Thom and Swart's (1940) wing model at Re = 1, which achieved its maximum lift-to-drag ratio at 45 degrees. These numbers illustrate how boundary layers at increasingly low Reynolds numbers make shape and orientation details of wings increasingly less significant. Thus, as Reynolds numbers drop below 10^3 wings become increasingly inefficient, and lift-based locomotion is simply not a viable option at Reynolds numbers much below 100.

Despite this low-Reynolds number effect, many insects do appear to operate in the Reynolds number range of 200–2000 (Vogel, 1994, p. 249). Weis-Fogh and Jensen (Jensen, 1956; Weis-Fogh and Jensen, 1956) used a *quasisteady* approximation—treating many instances during the stroke as separate steady states and then summing them up over the whole wing stroke—to show that standard airplane aerodynamics was adequate to explain lift production in the forward flight of locusts. Later researchers, however, discovered that conventional aerodynamics could not account for the amount of lift required in extreme situations, such as hovering (e.g., Norberg, 1975; Ellington, 1984b). The leading edge vortex, for example, allows a wing to operate at higher angles of attack than would a bound vortex, thus increasing maximum lift. The leading edge vortex also forms more quickly than the bound vortex, an important advantage for a bee that must generate a new vortex at the beginning of every downstroke, 100 times per second.

---

j. Thom and Swart (1940) do not give lift coefficient values for their model at Re = 1000, but by extrapolating from their graph, I estimate *L/D* to be approximately 6 at Re = 1000, for an RAF6 airfoil that in technical use (Re > 10^6) probably had a lift-to-drag ratio ≥10.

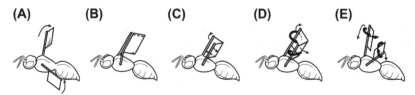

FIGURE 3.15 The clap-fling process discovered by Weis-Fogh. The sequence starts at the middle of the upstroke (A). At the top of the upstroke (B), the wings are "clapped" together. At the beginning of the downstroke, the leading edges peel apart while the trailing edges stay together (C). This causes air to rush around the leading edges, forming a vortex at the front of the wing (D). As the wings move apart, these vortices become the bound or leading edge vortices (and each other's starting vortices), producing lift almost immediately (E). *Modified from Alexander, D.E., 2002. Nature's Flyers: Birds, Insects, and the Biomechanics of Flight. Johns Hopkins University Press, Baltimore, Maryland, 358 pp., originally drawn by Barbara Heyford from data in Weis-Fogh, T., 1975. Flapping flight and power in birds and insects, conventional and novel mechanisms. In: Wu, T.Y., Brokaw, C.J., Brennan, C. (Eds.), Swimming and Flying in Nature, vol. 2. Plenum Press, New York, pp. 729–762.*

Perhaps the best-known mechanism of the unconventional effects used by animals is the *clap-fling mechanism*, first described in very tiny insects by Weis-Fogh (1973) and subjected to considerable analysis since then (e.g., Lighthill, 1973; Sane, 2003; Kolomenskiy et al., 2011). Insects that use this mechanism "clap" their wings together at the top of the upstroke (Fig. 3.15). To begin the downstroke, they peel the leading edges apart while the trailing edges stay together (the "fling"), like opening the pages of a book. The leading edge vortex forms as the leading edges move apart; it is thus already fully formed as the trailing edges start to separate as the wings move downward in the downstroke. Without clap-fling, a wing would have to move three or four chord lengths through the air at the beginning of the downstroke before the bound vortex is fully formed (see "Wagner effect," Alexander, 2002, p. 94), and at very low Reynolds numbers, it might need to move even farther. We will return to the clap-fling effect when we look at other unsteady effects at the end of this chapter.

3.5 SWIMMING

One fundamental difference between swimming in water and flying in air is that in air, the flyer's body weight must be supported as part of the locomotion mechanism. In water, due to the fact that animals are normally of a similar density to water, buoyancy from displacing water means that little or no locomotory effort need go into supporting body weight. Swimming locomotion is thus all about thrust production.

Another important difference is that the density and viscosity values of air versus water mean that, all else being equal, Reynolds numbers in water will be about 15 times higher than in air at room temperature, 20°C (the difference is strongly temperature dependent, ranging from 7.4 times at 0°C to 25 times at 40°C). Or to put it in another way, the Reynolds numbers—and hence the lift and drag coefficients—will be equivalent for the same object in a water flow and in an airflow 15 times faster. Note that equivalence of Reynolds numbers means equivalence of flow patterns and force coefficients, but *not* equivalence of forces. For example, an object moved from air to water at equivalent Reynolds numbers (at 20°C) will experience approximately 3.5 times greater forces in water.

3.5.1 Swimming Modes

Swimming animals can use lift-based mechanisms, drag-based mechanisms, or jetting to produce thrust. Because wings lose effectiveness at low Reynolds numbers, lift-based mechanisms are largely used by animals with bodies or appendages that operate at decently high Reynolds numbers. Drag-based mechanisms can work at any Reynolds numbers; at higher Reynolds numbers, they have advantages and disadvantages compared with lift-based methods, but for small animals operating at low Reynolds numbers, drag-based thrust may be the only reasonable option. Jet propulsion, as used, for example, by squid, is a fairly specialized mode generally limited to situations where efficiency is not a constraint; for long distance locomotion, they apparently depend more on undulating fins (see Section 3.5.4).

3.5.2 Lift-Based Swimming

Many medium-sized and large animals swim with pairs of airfoil-shaped appendages using a birdlike stroke, albeit modified so upstroke and down-stroke are more symmetrical like the porpoise mentioned above, including sea lions, sea turtles, and penguins. Others, such as cetaceans, tunas, swordfish, and some sharks, use the caudal "tail" fin as a hydrofoil. Those animals generally have stiff bodies connected by a narrow "peduncle" to a high aspect ratio, lunate (crescent-shaped) caudal fin. Some fish swim by flapping fins other than the caudal fin. Skates and rays have massively enlarged pectoral fins, and certain rays such as eagle rays and manta rays flap their pectoral fins in a very birdlike fashion. Triggerfish are stiff-bodied, laterally compressed fish with enlarged dorsal (top) and ventral (bottom) fins. They swim by flapping the dorsal and ventral fins right and left simultaneously, looking strikingly like a bird tipped up on its side, flapping stubby wings.

3.5.3 Drag-Based Swimming

Most really small swimmers, and a fair variety of larger ones, use drag-based propulsion. The concept is simple: move an appendage backwards in a high-drag configuration—broadside to the flow—and then move it forward in a low-drag configuration—edge-on to the flow. Many aquatic arthropods (insects, crustaceans) have appendages with a fan of bristles on pivoting joints. When the appendage moves backwards, the bristles fan out and lock perpendicular to the flow, and, depending on the density of bristles and the Reynolds number, the water acts as if they form a more or less solid surface producing flat-plate drag. During the recovery stroke, as the appendage swings forward, the bristles fold down parallel to the flow, producing much less drag (Fig. 3.16) (Nachtigall, 1980; Koehl, 1995). Many animals that need to walk on land as well as swim—muskrats, ducks—use webbed feet to do the same thing: the toes spread out and form a broad paddle on the back (power) stroke, and then squeeze together into a tight bundle for the forward (recovery) stroke.

Vogel performed a perceptive thought experiment to compare the characteristics of lift- and drag-based locomotion (Vogel, 1994, pp. 283–286). He described a set of appendages that, with modest changes in stroke pattern and orientation, could be used to produce thrust with either lift or drag, operating at a Reynolds number of approximately 10^4. He showed that when used as paddles (for drag), they gave high initial acceleration, but their top speed was

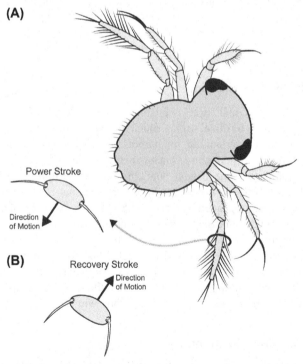

FIGURE 3.16 Folding bristles on the limbs of swimming arthropods, such as this juvenile water boatman. (A) Bristles fan out to increase area during the power stroke. (B) Bristles fold down to reduce area during the recovery stroke. *Artist: Sara Taliaferro.*

exactly as fast as they could be moved backwards. When used as hydrofoils, they produced much lower initial acceleration, but much higher top speed—five times higher in his example. Moreover, lift-based swimming is more energetically efficient at Reynolds numbers where airfoils are effective. For a muskrat or freshwater turtle that must walk on land as well as swim, the cost of drag-based swimming evidently is outweighed by the problems of trying to walk on a hydrofoil-shaped appendage; sea turtles with their hydrofoil flippers are exceedingly poor walkers.

On the other hand, for emergency escape behavior, where acceleration might matter a lot more than energy conservation, a paddle can be highly effective. Crustaceans such as crayfish and shrimp can flex their abdomens sharply, swinging their extended tail fan abruptly forward and sending the animal shooting backward (Daniel and Meyhofer, 1989). In fact, when you eat peeled shrimp or lobster tails, what you eat is muscle devoted almost entirely to this emergency "tail flip" escape behavior. Even fish that otherwise depend on lift-based propulsion sometimes use a drag-based behavior called the "C-start" for rapid acceleration from rest. The fish bends its entire body into a C-shape, and when it straightens its body, its initial acceleration is powered by drag on the sideways-moving tail (Webb, 1976). Scientists have measured accelerations of 5–25 times the acceleration of gravity during C-starts in a variety of fish (e.g., Harper and Blake, 1990).

3.5.4 Undulatory Swimming

Many swimmers with long, flexible bodies, such as eels, sea snakes, leeches, and some polychaete worms, use *undulatory* swimming. In undulatory swimming, waves of bending pass down the animal's body from front to back and increase in amplitude as they pass down the body (Fig. 3.17). Undulation does not fit neatly into the lift-based or drag-based categories. The traditional analysis, elongate slender body theory (Lighthill, 1971), is based on accelerational processes (see Section 3.7) and neglects viscosity and so is termed a "reactive" model (Daniel et al., 1992). On the other hand, because important forces, especially on the tail, operate at right angles to body motions, and because the pattern of water flowing off the tail fin's trailing edge seem to dominate thrust production, undulatory swimming at moderate to high Reynolds numbers can be considered a lift-based mechanism (Wu, 1971; Daniel, 1984; Vogel, 1994, 2013). Wardle and Videler (1980) give a nice comparison of the different approaches.

The hydrodynamics of undulation are not as well understood as wing physics, and the details of how they generate thrust are beyond the scope of this book. One interpretation, where the water acts as if a bound vortex forms on the outside of each bend as it travels down the length of the body and is then shed into the wake (Vogel, 1994, p. 281), is now considered an oversimplification. The tails of undulatory swimmers do shed vortices into the

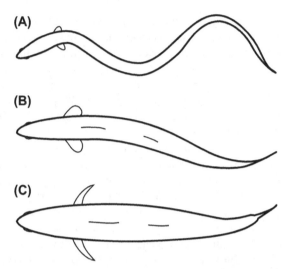

FIGURE 3.17 Ichthyological terminology for fish swimming modes. (A) Anguilliform (after *Anguilla*, the genus containing common eels), where the body at any instant forms at least one full wave of bending (e.g., eels). (B) Carangiform (after *Caranx*, the genus containing bar jacks), where the body forms less than a full wave (e.g., trout). (C) Thunniform (after *Thunnus*, a genus containing tuna), where the body is stiff and only the most posterior part bends (e.g., tunas).

wake at the end of each half stroke,[k] injecting momentum into the wake, with thrust as the equal and opposite reaction. The details of the mechanism that generates such vortices and the exact forces involved, however, are the subject of much active research (Lauder and Tytell, 2006; Lauder, 2015).

Some animals swim by undulating appendages rather than the whole body. The bowfin (*Amia calva*) is a long fish with a long dorsal fin running from just behind the head to the tail. Bowfins can swim forward or backward by sending undulations along their dorsal fin, either from front to back or from back to front. The various gymnotiform knifefishes often seen in pet stores do the same thing with their long ventral fin. Other animals swim by undulating paired lateral fins, including some skates, stingrays, squid, and cuttlefish.

Many fish swim with a motion that is intermediate between the sinuous full-body undulations of eels and the stiff-bodied tail-only beats of tuna. This intermediate mode, so-called *carangiform* swimming, is typical of generalized fish such as trout, bass, perch, etc. The whole body bends, but the front of the body is stiffer than an eel and it forms less than one full wave at any given time (Fig. 3.17B). Nevertheless, despite the traditional ichthyological distinction, in hydrodynamic terms carangiform swimming appears to be just

k. This vortex shedding is very similar to the shedding of the bound or leading-edge vortex at the end of the downstroke by wings of small or slow flying animals.

a form of undulatory ("anguilliform") swimming (Lauder, 2015). Moreover, evidence from flow visualization studies suggest that flow around fins anterior to the tail (especially the dorsal and anal fins) as well as asymmetrical movements of the tail fin itself, may be important in thrust production (Lauder and Tytell, 2006).

3.5.5 Swimming by Jetting

Jetting locomotion is perhaps the clearest example in nature of Newton's third law: eject a mass of water in one direction, the animal moves in the opposite direction. Unlike technological jet propulsion systems, in nature, jetting is usually pulsatile rather than continuous. Such a system is intrinsically simple, only requiring a body cavity with an outlet, surrounded by muscle. Medusan jellyfish, among the structurally simplest of animals, swim using pulsatile jet locomotion. If an animal can expel water fast enough, jetting can produce great acceleration. The Achilles heel of jetting is that to achieve useable accelerations and speeds, the water must be expelled through a small aperture at high speeds. One version of the Froude propulsion efficiency, η_F, is given by

$$\eta_F = \frac{2v_1}{v_2 + v_1} \tag{3.14}$$

where v_1 is the fluid speed before it is accelerated and v_2 is its speed after it has been accelerated; v_1 is normally taken as the animal's swimming speed. (Froude efficiency is based on the "actuator disk," a simplified description for airplane propellers and helicopter rotors; although animal locomotion may violate some of its assumptions, it gives important general relationships and it remains the conventional yardstick for swimming and flying efficiency (Vogel, 1994, p. 237, 2013, p. 114).) Eq. (3.14) says that accelerating a lot of fluid a small amount is more efficient than accelerating a little fluid a lot. To achieve decent speeds, jetting animals such as squid eject water at high speed through a small orifice, thus adding a lot of acceleration to a relatively small volume of fluid. All else being equal, for moderate-sized animals, jetting is the least efficient form of swimming, lift-based modes are most efficient, and drag-based swimming is somewhere in between. Nevertheless, for escape behavior, where the difference between escaping and becoming someone else's meal may depend on acceleration, efficiency may take a backseat to an inefficient mechanism that generates high accelerations. (Note that for slow cruising, many squids use undulating fins instead of jetting).

3.5.6 Swimming at Low Reynolds Numbers

For animals swimming at Reynolds numbers below 10—planktonic crustaceans, the smallest aquatic insects—paddles are the only practical option.

Bristly appendages are common in this size range, and depending on the Reynolds number and bristle spacing they can function as solid paddles even while giving a visual impression of significant spaces between bristles (Zaret and Kerfoot, 1980; Koehl, 1995). The smallest animals and many protists swim using cilia or flagella which are even smaller, microscopic appendages, but they still depend on a drag-based mechanism. Cilia and flagella are cell organelles with the same internal structure, but cilia are short and present in huge numbers, whereas flagella are much longer and present either singly or very few. Cilia and flagella are cylindrical in cross section, and their operation depends on differences in drag on cylinders in different orientations. For example, at low Reynolds number, a cylinder 100 times longer than wide will have approximately 1.8 times more drag moving through the fluid with its long axis perpendicular to the flow than with its long axis parallel to the flow. Cilia make use of this difference by moving perpendicular to the water during the power stroke and parallel to the water during the recovery stroke. For a detailed description of the operation of cilia and flagella, see Vogel (1994, pp. 252–256).

3.6 INTERNAL FLOWS

When any fluid flows in a pipe, channel, or tube, the walls enclosing the fluid influence the flow pattern. Thanks to the no-slip condition, the friction between the walls and the fluid generate velocity gradients. Just as a river's current can be much slower close to the shore than out in the middle, the flow in any enclosed channel will be slow near the walls and fastest near the middle. The flow profile is a graphical representation of the flow velocities at any cross section of the channel. As fluid enters a pipe, such as from a larger reservoir, the flow profile starts out uniform across the channel, so-called *plug flow*. As the fluid moves down the length of the channel, the influence of the walls extends farther into the channel, much like a boundary layer growing with distance along a flat plate (Fig. 3.18A). Since the fluid close to the walls moves slower, by continuity, the fluid in the center of the channel will have to move faster than the original uniform flow speed. For laminar flow far from any entrance or bend, the flow profile is parabolic and the flow is called *fully developed*.[1] Internal flows in plants and animals are almost always at Reynolds numbers less than 100 and so are laminar. (Humans, being on the large end of the animal size scale, sometimes experience

---

1. The distance between the entrance of a tube and the location where the flow is fully developed is the "entrance length," which is of great significance in engineered systems such as plumbing. Because of their low Reynolds numbers, in organisms, entrance lengths of internal flows are rarely more than a diameter or two and thus rarely significant; see Vogel (1994, pp. 296–299) for further discussion.

(A)

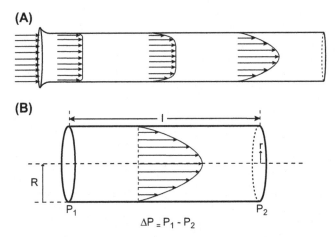

(B)

FIGURE 3.18 (A) Change of laminar flow profile in the entrance length of a tube, from plug flow at the entrance (left) to the fully developed, parabolic flow profile (right). Length of arrows indicates flow speed. (B) Pressure and flow in a cylinder with laminar, fully developed flow. l, length; P_1, pressure at location 1; P_2, pressure at location 2; r, distance from cylinder centerline; R, radius of cylinder; ΔP, pressure drop over length l.

turbulent blood flow at branches of the largest arteries at the highest flow speeds, or when disease narrows those arteries; larger animals may experience turbulent blood flows more routinely.) Moreover, for reasons of material economy and mechanical strength, many internal tubes and channels are circular in cross section, which allows us to use some simple equations. Finally, for flow in tubes, the principle of continuity does not allow the viscosity to slow the average flow speed along the length of the tube. The effect of viscosity, instead, is to reduce the pressure along the tube. Thus, the difference in pressure between an upstream location and a downstream location, ΔP, is what drives the flow.

The volume of fluid moving through a given area per unit time, or the *volume flow rate*, Q, is given by the Hagen—Poiseuille equation:

$$Q = \frac{\pi \Delta P R^4}{8\mu l} \tag{3.15}$$

where R is the tube radius and $\Delta P/l$ is the pressure drop per unit length of tube. Note that the flow is proportional to the radius to the fourth power, meaning that slight changes in tube radius can have overwhelming effects on volume flow rate. In contrast, volume flow rate is directly proportional to the pressure drop per unit length and inversely to the viscosity. Eq. (3.15) can also be used to define the *resistance* to flow (see Box 3.5).

For fully developed flow, the flow speed, v_r, at any distance r from the center of the tube is given by

---

BOX 3.5 Resistance to Flow

Resistance to fluid flow is analogous to resistance in an electrical circuit: it is the factor that determines how much push is needed to produce a given flow rate. For a fluid, resistance, R_s, is given by

$$R_s = \frac{\Delta P}{Q}.$$

For fully developed flow in a circular tube, we can substitute into the Hagen–Poiseuille equation and get

$$R_s = \Delta P \frac{8\mu l}{\pi \Delta P R^4} = \frac{8\mu l}{\pi R^4}.$$

This equation again illustrates that slight changes in a tube's radius can have quite disproportionate effects on resistance and hence volume flow rate. Fluid flow resistances combine exactly like electrical resistances: resistances in series are simply added, whereas for resistances in parallel, the sum of the reciprocals of the individual resistances equals the reciprocal of the total resistance.

$$v_r = \frac{\Delta P (R^2 - r^2)}{4\mu l} \tag{3.16}$$

which is the equation for the parabolic flow profile shown for fully developed flow in Fig. 3.18B. The maximum flow speed occurs in the center of the tube, so setting $r = 0$ gives the maximum flow speed, v_{max}:

$$v_{max} = \frac{\Delta P R^2}{4\mu l}. \tag{3.16a}$$

The average flow speed, $\bar{v}$, is the volume flow rate (Q) divided by the cross sectional area of the tube:

$$\bar{v} = \frac{\Delta P R^2}{8\mu l}. \tag{3.16b}$$

So Eqs. (3.16a) and (3.16b) yield the handy result that the maximum flow speed in a circular channel is exactly twice the average flow speed, which can be very convenient experimentally. Sometimes measuring outflow (e.g., time to fill a beaker of known volume) is easier, in other situations measuring the center flow speed (e.g., timing movement of a dye marker) is more convenient, but in either case, mean and maximum flow speed can easily be determined.

Although many transport tubes inside animals and plants have circular cross sections, many others do not. Thanks to the sensitivity of Q to changes in transverse dimensions, modest deviations from circularity will result in both the volume flow rate and flow speeds being different from those predicted by the above equations. Empirical corrections for "hydraulic diameter" are available to correct such discrepancies, although most such correction factors given in the engineering literature are for turbulent, rather than laminar, flow (Vogel, 2013, p. 165). Lewis (1992) gives correction factors for rectangular and oval tubes. Additionally, Vogel points out that for flow between flat plates much wider than the distance between them, the maximum flow speed is only 1.5 times the mean flow speed. This puts slightly higher flow speeds closer to the walls compared with tubes, which should improve material exchange between the fluid and the walls (Vogel, 2013, pp. 167–169). Many exchange surfaces can be modeled as channels between parallel plates (although the plates are not always flat), e.g., fish gills, spider book lungs, and the spiral valve of shark intestines.

Although exceedingly few internal systems in animals and plants are large enough to support turbulent flow, if it were present, turbulent flow could provide material or heat exchange benefits (at the cost of greater transport resistance). First, the flow profiles of turbulent flows are blunter (and mathematically much messier) than laminar flows, having maximum speeds much closer to the mean speed; in other words mean speeds near the walls are higher than those for laminar flows. Second, turbulent flows have substantial lateral transport, bringing fluid from the center of the channel close to the walls. Thus, if air became turbulent in elephant respiratory passages during vigorous exercise, the turbulence could help carry away the heat faster, albeit at an increased cost in moving the air. Similarly, turbulence in the respiratory airways of large dinosaurs would have influenced both the amount of effort needed to breathe and their ability to retain or give up metabolic heat and moisture.

3.7 WHEN FLOWS ARE NOT STEADY

The flow properties and relationships described so far are based on steady flows, i.e., flows that do not vary with time. For many engineering applications, flows are mostly steady. For flows of relevance to organisms, however, steady flows are more the exception than the rule. Consider the intermittent flow in your aorta caused by the pulsatile pumping of your heart, or the constantly changing flow patterns around a bird's wing as it flaps, or flow patterns around an undulating eel's body. Or consider the massive acceleration during the tail flip of a crayfish or shrimp. Biological flows can involve cyclic changes or large monotonic accelerations, and the equations we have seen so far in this chapter generally do not address either type of variation with time.

3.7.1 Continuous Acceleration: The Acceleration Reaction

When an animal's movement involves significant accelerations, the fluid imposes forces on the animal in addition to the drag forces. The effect is as if the animal is accelerating a certain volume of the fluid along with its own mass. Some fluid is, indeed, accelerating along with the animal, but not, unfortunately, a discrete or well-defined volume of fluid. Analyses using inviscid (ideal) fluid theory to model accelerations give surprisingly good agreement with accelerational forces in real fluids (Batchelor, 1967) and have given rise to the *acceleration reaction* concept (for a more detailed description, see Daniel, 1984). The acceleration reaction is a force on the animal's body in the *same* direction as drag for positive acceleration (increasing speed) and in the *opposite* direction from drag for negative acceleration or deceleration (decreasing speed). In other words, for positive acceleration, the acceleration reaction adds to drag, and for negative acceleration, it subtracts from drag. The acceleration reaction can be quantified by using the *added mass coefficient*, C_a, to calculate the additional force needed to accelerate, F_a:

$$F_a = C_a \rho V a \qquad (3.17a)$$

where ρ is fluid density, V is the organism's volume, and a is the animal's acceleration (note that ρV is the mass of fluid displaced by the animal's body). The total force, F_T, that an animal must overcome to accelerate in a fluid is thus

$$F_T = D + ma + C_a \rho V a \qquad (3.17b)$$

where D is total drag, ma is the force to overcome the animal's own inertia, and the third term is the force to overcome the fluid's acceleration.[m]

The added mass coefficient depends on shape and is lower for long objects (cigar shapes) moving with their long axes parallel to the flow and higher for flattened objects such as oblate spheroids moving broadside to the flow; spheres have a $C_a = 0.5$, and infinitely long circular cylinders perpendicular to the flow have $C_a = 1.0$. Streamlining should thus reduce the C_a as well as the C_D.

For high enough accelerations in water, the acceleration reaction can be considerable. Webb showed that for a crayfish using its tail flip to accelerate at 51 m s^{-2}, 90% of the fluid resistance was from the acceleration reaction and only 10% from conventional drag (Webb, 1979). The acceleration reaction turns out to be significant in the swimming of animals as varied as copepods and frogs (reviewed in Vogel, 1994, p. 365).

Although unsteady motion in air can produce biologically important effects, as we will see below, the acceleration reaction itself is hardly ever significant in air. Animals are so much denser than air that any added mass from accelerating air is insignificant; moreover, accelerations tend to be lower and speeds higher in air, so drag vastly outweighs any added-mass effects (Vogel, 1994, p. 368). From a biological perspective, the acceleration reaction

---

m. The sum of an accelerating animal's body mass and its added mass ($C_a \rho V$) is sometimes called its "virtual mass" (Daniel, 1984).

can thus be of great consequence for swimmers, but hardly ever for those living in air.

Whereas drag is generally proportional to surface area (length^2), the acceleration reaction is proportional to volume (length^3). One consequence is that, for C_a, we have no confusion about appropriate areas, we simply use the organism's volume. More significantly, the acceleration reaction grows disproportionately larger as animals increase in size: doubling an animal's length increases its acceleration reaction eightfold. Note that the acceleration reaction also acts on stationary (attached) organisms in accelerating flows, such as barnacles and mussels in wave-swept environments, although the governing equation changes slightly (Denny et al., 1985). The acceleration reaction may be the major factor limiting body size in such environments (Denny, 1988).

3.7.2 Unsteady Effects in Air

Unsteady effects in air normally occur in cyclic or oscillatory processes, rather than for simple acceleration. For example, cylinders in flowing fluids at Reynolds numbers ranging from 100 to 10^5 shed alternating vortices of opposite rotation, which form a "von Kármán vortex street" in the cylinder's wake. This causes a force on the cylinder perpendicular to the flow that reverses direction each time a vortex is shed. Such vortex shedding is why flags flap and taut wires hum in the wind. Other than avoiding the evolution of structures that have resonant frequencies near the frequency of vortex shedding, the shedding of von Kármán vortices seems to have little biological relevance (for details, see Vogel, 1994, pp. 369–373).

Most animal locomotion involves cyclic or oscillatory motion. In animal flight, for example, much research has gone into determining whether a quasisteady analysis is sufficient to describe the flows, or if the cyclic nature of flapping qualitatively changes the forces and flow patterns (Norberg, 1975, 1976a,b; Ellington, 1984b,c; Ennos, 1989). Lighthill (1975) developed the *reduced frequency*, a dimensionless index that indicates when unsteady effects may be significant. The reduced frequency, f_a, is given by

$$f_a = \frac{2\pi n c}{v} \tag{3.18}$$

where n is the flapping frequency, c is the wing chord, and v is the mean airspeed. At reduced frequencies greater than 0.5, unsteady effects become more likely. In forward flight, birds tend to operate well under 0.5, whereas some insects in forward flight have reduced frequencies at or near 0.5: locusts at 0.25 and fruit flies at 0.5; moreover, hovering animals tend to operate well above 0.5 (Vogel, 1994, p. 376). These values indicate that unsteady effects are not a major factor for birds in forward flight, but they may be for many insects (especially small ones) and unsteady effects may be dominant during hovering by all animals that can do so.

The Weis-Fogh "clap-fling" mechanism mentioned earlier is probably the best-known unsteady process in animal flight (Lighthill, 1973; Weis-Fogh, 1973). The bound vortex on a wing takes a finite amount of time to develop (the "Wagner effect"), and, depending on the Reynolds number, the wing may need to travel several chord lengths before the bound vortex is fully formed and full lift is being generated. For tiny insects such as mosquitoes and fruit flies that beat their wings several hundred times per second, this delay in lift production could put a significant dent in the total amount of lift they produce. Weis-Fogh observed that very tiny insects "clap" their wings together at the top of the upstroke, then they "fling" the leading edges apart while the trailing edges are still together, before the wings move apart in the downstroke (Fig. 3.15). As the leading edges fling open, they each form a bound vortex, or more likely, a leading edge vortex, as air rushes in past the leading edges to fill the widening gap between the wings. The wings do not have to shed a starting vortex—one of the processes that slows initial lift production—because each bound or leading edge vortex acts as the other wing's starting vortex (connected to each other by incipient tip vortices). Thus, lift production is essentially at full strength as the wings begin to move apart during the downstroke. Scientists have observed many medium and small insects using clap-fling (Ellington, 1984a), and even some birds such as pied flycatchers and pigeons seem to use it for brief bouts of hovering or vertical takeoffs (Weis-Fogh, 1975).

Researchers have proposed several other unsteady lift-enhancing mechanisms—wing flip (Weis-Fogh, 1975), delayed stall (Ellington, 1984a)—but so far, evidence is weak for animals using mechanisms other than clap-fling. Indeed, although organisms are more likely to experience unsteady than steady flows, the great difficulty of analyzing such flow patterns means that until recently, biomechanics researchers have relied largely on simplified approximations or empirically based relationships. With the continuing increase in the power of digital computers, researchers have begun to explore complex unsteady flows both experimentally—using computing-intensive techniques such as digital particle image velocimetry (e.g., Altshuler et al., 2009; von Busse et al., 2014; Han et al., 2015)—and analytically, using sophisticated computational fluid dynamics software (e.g., Hamdani and Naqvi, 2011; Mou et al., 2011; Shen and Sun, 2015).

FURTHER READING

General

Denny, M.W., 1993. Air and Water: The Biology and Physics of Life's Media. Princeton University Press, Princeton, NJ, 341 pp.

Vogel, S., 1994. Life in Moving Fluids: The Physical Biology of Flow, second ed. Princeton University Press, Princeton, NJ. 467 pp.

Flight

Alexander, D.E., 2002. Nature's Flyers: Birds, Insects, and the Biomechanics of Flight. Johns Hopkins University Press, Baltimore, Maryland, 358 pp.

Ellington, C.P., 1984. The aerodynamics of flapping animal flight. American Zoologist 24, 95–105.

Sane, S.P., 2003. The aerodynamics of insect flight. Journal of Experimental Biology 206, 4191–4208.

Tennekes, H., 2009. The Simple Science of Flight: From Insects to Jumbo Jets, Revised ed. MIT Press, Cambridge, Massachusetts. 201 pp.

Internal Flows

Caro, C.G., Pedley, T.J., Schroter, R.C., Seed, W.A., 2012. The Mechanics of the Circulation. Oxford University Press, Oxford, UK, 527 pp.

Lewis, A.M., 1992. Measuring the hydraulic diameter of a pore or conduit. American Journal of Botany 79, 1158–1161.

Reynolds, O., 1883. An experimental investigation of the circumstances which determine whether the motion of water shall be direct or sinuous, and the laws of resistance in parallel channels. Philosophical Transactions of the Royal Society of London 174, 935–982.

Swimming

Daniel, T.L., Jordan, C., Grunbaum, D., 1992. Hydromechanics of swimming. In: Alexander, R.M. (Ed.), Mechanics of Animal Locomotion. Springer-Verlag, New York, pp. 17–49.

Lauder, G.V., Tytell, E.D., 2006. Hydrodynamics of undulatory propulsion. In: Shadwick, R.E., Lauder, G.V. (Eds.), Fish Biomechanics. Academic Press (Elsevier), San Diego, California, pp. 425–468.

Webb, P.W., 1975. Hydrodynamics and energetics of fish propulsion. Bulletin of the Fisheries Research Board of Canada 190, 1–158.

Unsteady Processes

Daniel, T.L., 1984. Unsteady aspects of aquatic locomotion. American Zoologist 24, 121–134.

Hamdani, H.R., Naqvi, A., 2011. A study on the mechanism of high-lift generation by an insect wing in unsteady motion at small Reynolds number. International Journal for Numerical Methods in Fluids 67, 581–598.

Weis-Fogh, T., 1973. Quick estimates of flight fitness in hovering animals, including novel mechanisms for lift production. Journal of Experimental Biology 59, 169–230.

Chapter 4

Biological Materials Blur Boundaries

So far, we have treated solids and fluids as separate, nonoverlapping categories. We defined solids as materials that resist shear and fluids as materials that resist the rate of being sheared. Very few biological solids fit this distinction perfectly. Most biological solids have a significant time-dependent (or fluidlike) component in their behavior, and so they are called *viscoelastic* solids. Also, some biological fluids show disproportionate responses to increasing shear rate, some disproportionately small, a few disproportionately large; in other words, their viscosities vary with flow speed. Moreover, some biological materials, such as some kinds of mucus, can transition between a soft solid and a viscous liquid, depending on the amount of shear. This chapter will focus on types of materials that do not fit the standard solid and fluid categories we have seen so far.

4.1 VISCOELASTIC SOLIDS

A viscoelastic solid is one with properties that vary depending on the duration of the test or the rate of loading. When tested very rapidly, such a material gives very different stress and strain results than when tested slowly or when loaded and unloaded in rapid cycles. A purely elastic material, such as room-temperature steel below its yield strain, gives the same stress—strain results no matter how quickly or slowly it is stressed. A biological material, such as the collagen that makes up most of a tendon, gives quite different results if stressed over a fraction of a second versus over minutes or hours (Fig. 4.1) (e.g., Haut and Little, 1969). Moreover, if I hang a weight from a steel cable, the cable strains a fixed amount, whether the weight is there for a millisecond or a month. If I hang a weight from a fresh tendon, the strain gradually continues to increase, so that the strain reading is different after a second or an hour or a week.

4.1.1 Transient Tests

The tendon example above can be quantified by the conceptually simplest test for viscoelasticity, the *creep test*. In a creep test, a constant stress is applied to

Nature's Machines. http://dx.doi.org/10.1016/B978-0-12-804404-9.00004-9

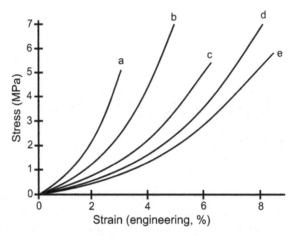

FIGURE 4.1 Stress—strain curves for a dog knee ligament (mostly collagen) measured in tension at different strain rates. (Note: only initial part of curves shown, not measured to failure.) Strain rates (% s^{-1}): a = 0.613; b = 0.255; c = 0.180; d = 0.0433; e = 0.0283. *Redrawn from data in Haut, R.C., Little, R.W., 1969. Rheological properties of canine anterior cruciate ligaments. Journal of Biomechanics 2, 289—292.*

the sample, and the strain is measured over time (Fig. 4.2A). The strain will continue to increase over time, more or less quickly depending on the material. When the stress is removed, many viscoelastic materials will recover to their original unstrained length, although recovery can take a very long time. Other materials only recover partway and exhibit some permanent deformation. Sea anemones, for example, can use their muscles to contract from tall columns to short lumps, and their reexpansion back to full height is at least partly, if not mostly (Vogel, 2013), driven by slow relaxation of the compression of its main body wall component, *mesoglea* (see Section 4.1.4.9 for details).

As the sample in a tensile creep test continues to extend over time, the strain at any instant can be used to calculate an elastic modulus[a] as a function of time, $E(t)$, which starts out high, as the initial or instantaneous modulus, E_i, and then falls off over time. These values of $E(t)$ can then be plotted against time, giving a curve like that of Fig. 4.3. The results of a creep test are usually summarized with the *retardation time*, τ', which is the time required for the modulus to drop from E_i to E_i/e (where e is the base of natural logarithms); τ' can be read off the curve in Fig. 4.3.

Another type of transient test of viscoelasticity is the *stress-relaxation test*. In a stress-relaxation test, the sample is abruptly stretched (or otherwise deformed) a fixed amount and the stress in the sample is measured over time (Fig. 4.2B). In this case, rather than the strain increasing, the stress decreases

---

a. Results of creep tests are often expressed as *compliance*, D, where $D(t) = \varepsilon(t)/\sigma$, which is simply the inverse of the elastic modulus.

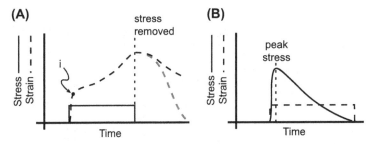

FIGURE 4.2 Transient tests for viscoelasticity. (A) A creep test applies a constant stress and monitors strain; the test ends when the stress is removed. Some viscoelastic materials eventually return to their original length (gray dashed curve), while others show some permanent deformation (black dashed curve). Dot at *i* is point of initial (maximum or instantaneous) elastic modulus, see Fig. 4.3. (B) A stress-relaxation test applies a constant or fixed strain and monitors the course of the decrease in stress from the peak stress value. *Modified from Alexander, D.E., 2016. The biomechanics of solids and fluids: the physics of life. European Journal of Physics 37, 053001,* © *European Physical Society. Reproduced by permission of IOP Publishing. All rights reserved.*

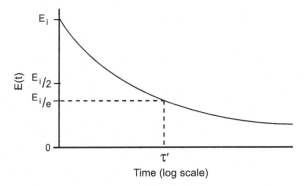

FIGURE 4.3 Modulus of elasticity, $E(t)$, as a function of time derived from a creep test. e, base of natural logarithms; E_i, initial elastic modulus, corresponding to i on Fig. 4.2; τ', retardation time.

over time. The characteristic value for a stress-relaxation test is the *relaxation time*, τ, which is the time for the stress (rather than the elastic modulus or compliance) to fall from its initial value, σ_i, to a value of σ_i/e. Using a derivation based on the conceptual equality of the shear modulus and viscosity (Vincent, 1990, p. 12), we find that for a viscoelastic material, $\tau = \mu/E$. In other words, the relaxation time gives the ratio of the viscosity to the elastic modulus (Wainwright et al., 1982, p. 25), a handy index of the relative liquidness or solidness of a viscoelastic material.

4.1.2 Springs and Dashpots

Scientists have devised various analytical models (Vincent, 1990, p. 16) and "spring and dashpot" models (Fig. 4.4) (Wainwright et al., 1982, pp. 33−39) to

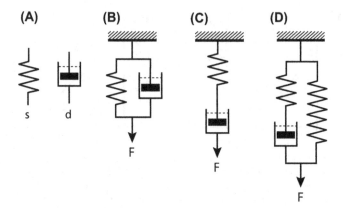

FIGURE 4.4 Spring and dashpot models of viscoelasticity. (A) Elements of model: a spring, *s*, represents pure elasticity with a particular modulus of elasticity and a dashpot, *d*, represents pure viscosity, with a given dynamic viscosity; elements can be added and combined until a model produces the same behavior as an actual material (usually limited to a particular range of stress and strain). (B) The Kelvin–Voigt model, a simple parallel model. (C) The Maxwell model, a simple series model. (D) The standard linear solid model, combining series and parallel elements. *F*, force.

characterize viscoelastic behavior, and these are regularly applied to biological materials (e.g., Alexander, 1962). As Vincent (1990) points out, these models assume linearly viscoelastic materials[b] at strains of less than 0.01, neither of which applies to typical biological materials as such materials are used in organisms. These models can provide a sophisticated description of the behavior of a viscoelastic material over a limited range of stress and strain but do not necessarily illuminate how such behavior is related to the underlying molecular processes (Vogel, 2013, p. 352). Moreover, in materials science, looking at a material's viscoelasticity over a wide temperature range is typically used so that high temperatures can substitute for very long test durations, allowing materials to be tested over the equivalent of several orders of magnitude in time. Because such high temperatures degrade biomaterials, they usually cannot be characterized over anywhere near the wide time ranges scientists commonly use for artificial polymers.

Both the creep test and the stress-relaxation tests are *transient* tests.

4.1.3 Dynamic Testing

A somewhat more sophisticated approach to characterizing viscoelasticity uses cyclic, repetitive loading and unloading to perform *dynamic testing*. Researchers normally use sinusoidal loading for mathematical and experimental

---

b. A linearly viscoelastic material has a Hookean elastic modulus and a Newtonian viscosity, which apparently applies to relatively few artificial polymers, and essentially no biomaterials.

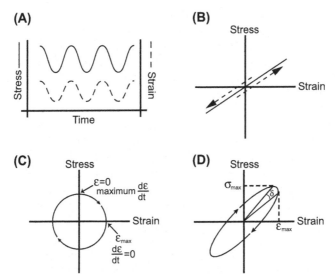

FIGURE 4.5 Results of dynamic stress—strain tests. (A) When an oscillating stress is applied to a purely elastic material, the strain will be exactly in phase with the stress. (B) For stress oscillating about zero (alternating tension and compression), the dynamic stress—strain curve for an elastic solid will be a single line that may or may not be perfectly straight; as the sample is compressed, values will travel down the line, and as the sample is pulled in tension, values will travel up the line. (C) A viscous liquid gives a dynamic stress—strain curve of a circle, because the maximum stress occurs at the maximum strain rate (at zero strain) and the minimum stress occurs at zero strain rate (at maximum strain). (D) The dynamic stress—strain curve for a viscoelastic material is an ellipse, with phase angle δ given by the angle between the points of maximum stress (σ_{max}) and maximum strain (ε_{max}) on the curve.

convenience, and they try to use a frequency or range of frequencies of functional relevance to the organism. (Biological materials tested at very different frequencies may produce different results.) If we look first at the results of a dynamic test of an elastic material and a viscous liquid, we can then see how the behavior of a viscoelastic material is in between.

In a dynamic test, a sinusoidal strain above and below zero (i.e., extending and compressing the material about zero) is applied to the test material. A perfectly elastic material will have its stress exactly in phase with its strain (Fig. 4.5A), so its dynamic stress—strain graph will be a more or less straight line; sample values travel up and down this line as the material is tested (Fig. 4.5B). Now consider a viscous liquid, perhaps a paddle with a sinusoidal oscillation in a vat of liquid (since a liquid cannot be clamped in a universal testing machine). If we apply a sinusoidal strain to the liquid, the stress will be 90 degrees out of phase because the strain *rate* will be near zero at maximum strain and the strain *rate* will be highest near zero strain (Ennos, 2012, p. 45). This gives a dynamic stress—strain curve of a perfect circle, because the stress and strain are exactly 90 degrees out of phase with each other (Fig. 4.5C). The strain of a viscoelastic material will also lag the stress, but by less than 90 degrees, producing an elliptical stress–strain curve (Fig. 4.5D). The area inside this ellipse represents the energy lost to viscosity: the wider the ellipse, the larger the viscous component of the viscoelasticity and the farther the material is from a purely elastic solid (Vogel, 2013, p. 355).

The phase angle, δ, measures the lag between strain and stress, and scientists use it as another way to quantify viscoelasticity. (Note that $\delta = 0$ degree for a purely elastic material and $\delta = 90$ degrees for purely viscous fluids.) The phase angle can be read from the dynamic stress–strain graph, where it is the angle between the points on the curve of maximum stress and maximum strain (Fig. 4.5D). This angle can also be calculated from the dynamic stress–strain equations. The strain at any given point on the curve is given by

$$\varepsilon = \varepsilon_0 \sin(\omega t) \tag{4.1}$$

from the definition of sinusoidal motion, where ε_0 is the maximum strain, ω is the angular velocity, and t is the time. The stress is given by

$$\sigma = \sigma_0 \sin(\omega t + \delta) \tag{4.2}$$

where σ_0 is the maximum stress, thus giving the phase angle, δ.

These relationships allow us to define two new moduli (Vogel, 2013, p. 356). The first is the real or elastic modulus, E':

$$E' = E_0 \cos\delta \tag{4.3}$$

(where E_0 is the maximum elastic modulus given by σ_0/ε_0). This elastic modulus is often called the *storage modulus* because it is a measure of how much strain energy is stored and released. The imaginary or viscous modulus, E'', is given by

$$E'' = E_0 \sin\delta \tag{4.4}$$

where E'' represents the energy that is lost as heat in the material, so it is also called the *loss modulus*. Finally, $\tan(\delta)$ is the viscoelastic *damping*, which relates E' and E'' and represents a relative measure of energy loss:

$$\tan\delta = \frac{E''}{E'}. \tag{4.5}$$

The relationships given above must be used with caution for biological materials. As Vincent (1990, p. 22) points out, these relationships assume linear viscoelasticity and typical biological materials are not linearly viscoelastic. Because nonlinearly viscoelastic materials may appear linearly viscoelastic over limited ranges of strain, dynamic testing of biomaterials should use relatively low maximum strains, and strain rates and maximum stresses should be chosen to mimic as closely as possible values that are biologically relevant.

4.1.4 Biological Examples of Viscoelastic Materials

4.1.4.1 Bone

Vogel (2013, p. 357) observed that softer, more pliant biomaterials tend to be more viscoelastic, implying that harder, more rigid biomaterials should be less viscoelastic. Vertebrate bone is a good example of this. Bone generally functions as a rigid support or hard protective covering, and up until this point we have treated it as an elastic solid. In fact, bone tends to have a slightly higher Young's modulus when tested very rapidly than when tested more slowly (Vogel, 2013, p. 357). Bone's viscoelasticity is, nevertheless, rather small: the tan(δ) for bone from human tibia in one study was at most 0.01 (Vincent, 1990, p. 192), meaning that the storage modulus, E', was at least 100 times larger than the loss modulus, E''. Though small, this amount of viscoelasticity is enough to improve such bone tissue's resistance to shock loading and cyclic loading—during running—at biologically relevant frequencies.

4.1.4.2 Wood

Wood, another fairly rigid biomaterial, is somewhat more viscoelastic than bone, at least in its natural state (in trees). Researchers have long known that the viscoelasticity of wood is strongly affected by its water content, but most research has been done on dried ("seasoned") lumber for building use. Placet et al. (2007) performed one of the very few studies of green wood, fresh-cut from trees. At biologically relevant temperatures, they measured tan(δ) values from approximately 0.05 to 0.10 for wood dynamically tested parallel to the grain, including both hardwoods and softwoods. Wood is thus considerably more viscoelastic than bone, although the elastic storage modulus is still at least 10 times higher than the viscous loss modulus. This amount of viscoelasticity could be helpful in damping the effects of shedding Von Kármán vortices in high winds or irregular gusts during storms.

4.1.4.3 Arteries

Arteries of vertebrates and cephalopods, especially large ones near the heart, are well known for being stretchy. In mammals, they contain a substantial amount of elastin (in addition to lots of collagen), and their stretchiness serves an important function. When a mammal or octopus heart contracts, it forces a pulse of blood into the largest artery. By being stretchy, the artery can inflate slightly to absorb some of the volume of the pulse. Then when the heart relaxes between contractions, the artery recoils elastically and pushes blood away from the heart (Fig. 4.6). Thus the large arteries maintain blood flow between heartbeats and help smooth out the flow speeds: in the smallest arteries ("arterioles"), blood flow speed is essentially constant with no sign of pulses from the heart (Shadwick and Gosline, 1985, 1995; Withers, 1992, p. 674).

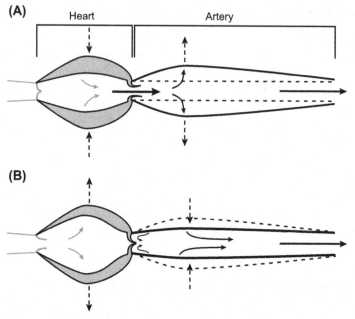

(A)

Heart   Artery

(B)

FIGURE 4.6 Artery elasticity smoothes blood flow. (A) Schematic cross section of heart (simplified to a single chamber) and artery. While the heart contracts, the outlet valves (right-hand side of heart) are open, and the increased pressure from the blood being forced out of the heart causes the artery to stretch and inflate. (B) Between contractions, the heart relaxes, allowing the outlet valves to close. Elastic recoil of artery causes artery to deflate, maintaining blood flow between heart's contractions. Dashed arrows, movement of heart or artery walls; solid arrows, movement of blood.

This function of the large arteries would seem to put a premium on elasticity, yet studies show that these arteries have significant viscoelastic behavior. For example, dog aortas had tan(δ) values between 0.1 and 0.2, meaning a viscous loss component of as much as 20% of the elastic component (Bergel, 1961). Such a high tan(δ) would seem to indicate significant loss of energy from the pumping work done by the heart. Shadwick (1999) suggested that this viscous component of the arteries' properties might be important in attenuating the pressure wave from each heartbeat that propagates at near-sonic speed through the arterial system, and doing so might be especially important for minimizing any possible resonances that might place unacceptable loads on the vessels.

4.1.4.4 Cartilage

Cartilage is a moderately stiff, flexible tissue most commonly found forming the bearing surfaces where bones come in contact with each other at joints. Cartilage provides both a smooth, low-friction surface and a shock-damping cushion. Cartilage's viscoelastic behavior plays a significant role in its cushioning ability.

Cartilage consists of a three-dimensional meshwork of collagen fibers in a gel matrix of proteoglycans (a type of protein) and water. Cartilage has long been known to have a time-dependent component to its mechanical behavior, but we now know that at least part of cartilage's viscoelasticity is not due to its molecular arrangement but to movement of water through pores in the solid components (Vincent, 1990, p. 124). Regardless of the mechanism, dynamic testing of cartilage from humans, pigs, and cows shows that for rapid loading and unloading—relevant for walking and running—it has tan(δ) values ranging from 0.18 to 0.23 (Ronken et al., 2012; Espino et al., 2014). Thus, the loss modulus is about one-fifth the storage modulus, which aids shock absorption. Curiously, at longer timescales, tens of seconds rather than fractions of a second, tan(δ) takes a big jump, up to 0.7; in this situation, the loss modulus approaches the magnitude of the storage modulus, due to water being squeezed out of the loaded region (Ronken et al., 2012). This slow creep at long timescales helps explain the well-known phenomenon that human adults are measurably taller right after getting out of bed in the morning than at the end of the day. Over the course of the day, water is squeezed out of the cartilage of our intervertebral disks (separating the vertebrae of our spine) due to carrying the body's weight, thus allowing our vertebral column to shorten. When we sleep, the disks are unloaded and expand by taking up water osmotically to replace that squeezed out (Ennos, 2012, p. 112).

4.1.4.5 Skin

In Chapter 2, we noted the distinctly J-shaped stress—strain curve for mammalian skin. Skin is also viscoelastic, and the degree of viscoelasticity can vary substantially with species, age, and body region. For example, the skin on human fingertips—tested in life using a noninvasive dynamic mechanism—gave a tan(δ) range from 0.055 to 0.13. Even though the viscous component was never more than about one-eighth of the elastic component, the difference between patches of skin with the lower tan(δ) and higher tan(δ) values was enough to significantly affect the sensitivity of particular sensory nerve endings (Yildiz and Güçül, 2013). In contrast, skin from cow bellies had tan(δ) values ranging from approximately 0.2 to 0.45, although the functional significance of this difference, if any, remains unclear (Ventre et al., 2009). Just from casual observation of skin properties, I suspect both the instantaneous and time-dependent mechanical properties of skin vary widely both across body regions and from individual to individual, even within a species. One of my elderly relatives merely needs to bump a hard object with her shin to tear the skin open, whereas a young person bumping just as hard might not suffer any more damage than brief pain.

4.1.4.6 Insect Cuticle

An insect's exoskeleton (outer body covering) is made up of cuticle, a complex layered composite with perhaps the widest range of material properties of any single class of solid. It ranges from very compliant, such as the flexible cuticle between abdominal plates, with Young's moduli in the range of 1–60 MPa (Reynolds, 1975), to the cutting edge of locust mandibles with a Young's modulus of 15,000 MPa (Cribb et al., 2008).

Beetles use modified front wings called *elytra* as protective covers for the hind wings and dorsal abdomen. Lomakin et al. (2011) made dynamic tests on the elytra of one species of beetle and found $\tan(\delta)$ of approximately 0.08, indicating relatively little viscous behavior for this stiff ($E = 5.8$ GPa) form of cuticle. On the other hand, many insects need to expand their abdomens considerably, for example, to accommodate large blood meals or to extend the abdomen for egg-laying. Reynolds (1975) found that the Young's modulus of the cuticle from the expandable abdomen of a blood-sucking true bug dropped from 62 MPa before a meal to 2.5 MPa after a meal (Fig. 4.7A; mosquitoes appear to have similar properties and show the expanded abdomen more clearly); the postfeeding cuticle showed considerably more creep and stress-relaxation, but because of the way the data were presented, τ or τ' could not be calculated (and no dynamic tests were performed so no $\tan(\delta)$ is available). Similarly, Vincent and Wood (1972) described cuticle from the abdomen of a female locust that greatly extends its abdomen for egg-laying, laying its eggs much deeper in the soil than the total length of its abdomen (Fig. 4.7B). The digging appendage at the end of the abdomen pulls on the abdomen just hard enough to take advantage of a stress-softening behavior of the intersegmental cuticle (flexible cuticle between hard plates), allowing it to stretch to strains of approximately 15 ($\sim 1500\%$)! Again, no dynamic tests were done on these locusts, but the enormous strains were fully reversible. Special muscles contract to shorten the abdomen immediately after egg-laying, but this is just a way to speed up the recoil, because the muscles soon relax and the abdomen stays shortened until the next bout of egg-laying a week or two later.

4.1.4.7 Keratin in Hooves

Keratin is an important structural protein, found in mammal hair, bird feathers, and the outer layer of mammalian and reptilian skin, rhinoceros horns, and hooves and claws. Horse hooves are moderately viscoelastic, with $\tan(\delta)$ values averaging around 0.145. The exact function of hoof viscoelasticity is not well understood. Hooves are stiffer and tougher at higher strain rates, apparently allowing the hooves to resist impacts as the horse increases its speed of locomotion. At lower speeds, the hooves show more marked viscoelasticity, perhaps providing some shock absorption (Kasapi and Gosline, 1996). Given the significant level of viscoelasticity, one might wonder about the consequences of creep in the hooves of animals that are on their feet almost continuously, but to my knowledge, this has yet to be investigated.

(A)

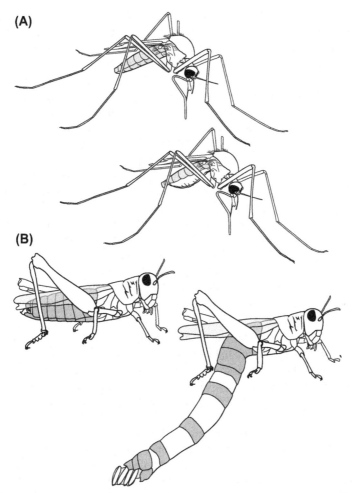

(B)

FIGURE 4.7 Changes in modulus of elasticity of insect abdominal cuticle. (A) Reduction in the elastic modulus of the cuticle of blood feeders, such as this mosquito, allows the abdomen to swell with a blood meal. (B) Stress softening of the intersegmental cuticle of the abdomen allows some grasshoppers to greatly extend the abdomen for egg-laying in the soil. Shaded regions of abdomen are hard sclerites, whereas intersegmental cuticle between sclerites is unshaded. *Artist: Sara Taliaferro.*

4.1.4.8 Spider Silk

The viscoelasticity of spider silk has important functional consequences. Spiders produce several types of silk. "Major ampullate" silk is the silk used for the main framework of orb webs, as well as the dragline that spiders use as a safety line and for "rappelling" down from high perches. As shown in Table 2.1, this type of silk has both high strength and high extensibility, giving it a toughness on par with steel. We saw in Chapter 2 that spider silk also has rather low resilience. That low resilience is reflected in the tan(δ) for this silk of 0.17 (Blackledge and Hayashi, 2006), meaning the loss modulus is about one-sixth of the storage modulus. In other words, one-sixth of the energy put into stretching the silk is lost as heat. Thus, the frames of orb webs help absorb the shock of insects impacting the web, absorbing the energy rather than releasing it and springing them back out of the web. Also, spiders apparently use draglines with strengths only slightly higher than their weight (low "factors of safety"), so the loss modulus helps cushion shocks during falls, as well as reducing the "bungee jump" recoil that might otherwise occur.

Spider silk from the sticky capture spiral thread of an orb web is even more specialized. It has a low elastic modulus but moderate strength and extreme extensibility ($\sim$170%), giving it very high toughness. Moreover, its storage modulus was so low, researchers could not measure it until the silk had extended over 100%, then it showed a tan(δ) similar to major ampullate silk (Blackledge and Hayashi, 2006). The great extension, low resilience, and significant loss modulus of the capture silk help cushion and absorb the impact of insects and prevent them from bouncing off the web.

4.1.4.9 Sea Anemone Mesoglea

The mesoglea of sea anemones is one of the best-studied viscoelastic biological materials in the animal kingdom. Mesoglea is a nonliving, secreted material that makes up most of the thickness of the sea anemone body wall. It is a soft, highly extensible, gel-like material consisting of random, scattered collagen fibers ($\sim$7%) embedded in a gel matrix of inorganic salts (5%), protein-polysaccharide complex (2% or 3%), and the rest, approximately 86% water (Gosline, 1971a). The mesoglea of *Metridium senile*, a large (40–50 cm tall) sea anemone about four times taller than wide, can tolerate strains of 2 or 3 without breaking, so clearly the matrix dominates its material properties and the collagen fibers are discontinuous and not cross-linked (Vincent, 1990). The collagen fibers probably act to increase toughness and as filler to slightly increase stiffness. The mesoglea's instantaneous elastic modulus was approximately 0.1 MPa, but in a stress-relaxation test, it fell by a factor of 100 over a duration of 10^5 s (a bit longer than a full day) (Gosline, 1971b). At low frequencies (0.1 Hz), the tan(δ) was just under 0.4, increasing to the range of 0.5–0.8 at higher frequencies (1.0 Hz), so at most frequencies it had a loss modulus almost as high or higher than the storage modulus. Results of creep tests gave a retardation time of 1–2 h (Alexander, 1962; Vincent, 1990). The consequences for the animal are that at frequencies in the order of a fraction of a second—such as the duration of a passing wave—the mesoglea was fairly stiff but still flexible. Over much longer intervals, such as durations that might represent a steady tidal current, it was less stiff, allowing the animal to passively bend into a lower-drag shape. The retardation time approximately

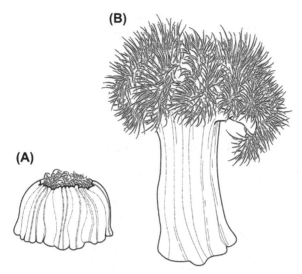

FIGURE 4.8 The sea anemone, *Metridium senile*. (A) Mostly contracted; if fully contracted, it would be less than half as tall. (B) Expanded to its full height of approximately 50 cm. *Artist: Sara Taliaferro.*

matches the time needed to reinflate after contracting (Fig. 4.8). Muscles power contraction of the body, and the general textbook explanation says that ciliary pumping drives expansion; but given the pressure limitations of ciliary pumping, Vogel (2013, pp. 358–359) argued that the slow recoil of the mesoglea's strain (with a time constant in the order of an hour) seems more likely to power reinflation.

4.2 NON-NEWTONIAN LIQUIDS

A Newtonian fluid is one whose viscosity is not affected by shear rate: all else being equal, flow speeds or shear rates do not change the viscosity. Air and water are both Newtonian fluids. Some liquids,[c] however, have viscosities that change with rate of shear. The two basic categories are *shear thickening* and *shear thinning*, and the names are fairly self-explanatory. The viscosity of a shear-thinning liquid decreases with shear—as it flows faster through channels or over obstacles, its viscosity drops. Many commercial house paints, quicksand, and whole blood are shear thinning. Shear-thickening liquids do the opposite—as shear rates and flow speeds increase, their viscosities increase. Cornstarch slurries and, under some conditions, the synovial fluid that lubricates our joints are shear-thickening liquids.

---

c. I am not aware of any non-Newtonian gases. Whether or not non-Newtonian gases exist, all gases of biological relevance are Newtonian.

Fluids can also be non-Newtonian with respect to time as well as to shear, somewhat analogous to creep in a viscoelastic solid. Fluids with a viscosity that decreases over time are called thixotropic, whereas those whose viscosity increases over time are called rheopectic. Some biological fluids may be thixotropic or rheopectic, but studies on time-dependent behavior of the viscosity of biological fluids seem to be lacking aside from anecdotal observations that some organisms live in mud or silt that is thixotropic.

4.2.1 Non-Newtonian Behavior of Everyday Liquids

Non-Newtonian fluids can have unexpected, even bizarre, behavior. Most shampoos and some polymer suspensions are shear thinning and display a startling behavior known as the Kaye effect (commonly called "looping," "leaping," or "jetting") (Versluis et al., 2006). If you pour a thin stream of shampoo into a container partly full of shampoo, a lump of shampoo builds up at the base of the stream due to viscosity. After the lump has built up sufficiently, a thin stream of shampoo will shoot out of the side of the lump, often forming a lassolike loop and rearing up, sometimes vertically, before collapsing after a second or so. (An Internet search for "Kaye effect shampoo videos" is well worth the effort.) Versluis et al. (2006) analyzed the physics of looping and concluded that it is entirely due to the shear-thinning viscosity of the shampoo.

Shear-thickening liquids can also show surprising behavior. A slurry of cornstarch in water, about the consistency of heavy cream, is a favorite teaching tool of biomechanics instructors. When poured or stirred slowly, it acts just like any viscous liquid. When stirred rapidly, the mixture suddenly acts like a soft solid, greatly resisting the stirring motion and even fracturing into chunks, which immediately slump back into liquid form as soon as stirring stops. If you gently lay a large spoon on the surface of the slurry, it sinks in, but if you tap the spoon sharply on the mixture, the spoon rebounds as if the mixture were solid, and the mixture may even show cracks briefly. If you stir near the bottom or sides of the container, it feels as if the mixture has a thickened layer close to these solid surfaces, but this again is an effect of increasing shear rates, and no separate thickened layer exists.

4.2.2 Blood

Whole mammalian blood shows slight shear thinning at normal concentrations of red blood cells (about 45% by volume in normal human blood). This effect is mainly because, at low shear rates, red cells clump together into long strands due to charge attraction and cell surface protein interactions. The liquid fraction of the blood, *plasma*, is Newtonian and has a dynamic viscosity about twice that of water. The long strands of red blood cells, however, increase the viscosity of whole blood to about four times that of water at normal body temperatures (Downey, 2003). In the large vessels, as the flow speed increases, the viscosity decreases as shearing breaks the red cell chains into shorter and shorter chains and eventually to individual cells. This particular effect is modest but could slightly reduce the amount of energy needed to pump blood during exercise, when flow speeds in large vessels are highest (Vogel, 2013, p. 350).

Because whole blood is not actually a continuum but a suspension of red blood cells in a liquid plasma, the physical behavior of blood changes as blood vessel diameters approach the dimensions of red blood cells—approximately 8 μm in humans. In fact, as blood vessels decrease in diameter from 1000 μm down to 3 μm, the blood's viscosity at first decreases down to a vessel diameter of approximately 10 μm, then increases sharply, by a factor of 3 or 4, from vessel diameters of 10 μm down to 3 μm (Secomb, 1995). Note that capillaries, the smallest vessels, are often narrower than the resting diameter of a red blood cell, so the red blood cells must deform to fit through the smallest vessels. The flow of blood in small vessels is surprisingly complex, with cells forming a near-solid stack in the center ("axial streaking") and leaving a zone of pure plasma close to the vessel walls. Even though the viscosity increases significantly in the smallest vessels, both theoretical and empirical studies show that the effective viscosity would be even higher if the cells filled more of the cell-free, plasma-only volume near the walls (Secomb, 1995).

The mechanics of blood flow, including the physical behavior of blood, appears to be the most intensively studied topic in all of biomechanics. To locate that research, one should know that *rheology* is the study of the flow of deformable materials including viscous liquids; much of the research literature on blood flow is thus found under the topic of "biorheology." For a detailed look at the mechanics of blood flow in the smallest vessels, see Secomb (1995). For a recent overview of the mechanics of circulation in general, see Caro et al. (2012).

4.2.3 Synovial Fluid

To move, animals usually use muscles to deform or change the arrangement of some sort of supporting skeleton. In the case of vertebrate animals, muscles move bones relative to one another, and these skeletal movements occur at articulations (joints). Most such articulations are *synovial joints*, the so-called "freely movable articulations." Where the bones come in contact at a synovial joint, they are covered with a layer of smooth cartilage forming the contact or *bearing* surface. Tissues surrounding the joint secrete a liquid called *synovial fluid* into the joint where it lubricates the bearing surfaces. Synovial fluid is a fairly dilute (3% or 4%) solution of hyaluronic acid (a long-chain polysaccharide polymer) along with some protein, in water.

The physical behavior of synovial fluid is so complex that some have described it as shear thickening (Vogel, 2013, p. 350) while others have called it shear thinning (Vincent, 1990, p. 88). In fact, synovial fluid is shear thinning under some conditions and shear thickening under others. For example, one early study used variations on standard engineering methods of measuring oil viscosity and concluded that at low shear rates, synovial fluid's viscosity is much higher than the thickest machine oil, but fell with increasing shear rate by a factor of more than 5 (Barnett, 1958). Other studies, using a version of dynamic testing suitable for liquids, found that at very high shear rates, synovial fluid (or equivalent solutions of hyaluronic acid) acts like a soft, elastic solid (Gibbs et al., 1968), which could be seen as dramatic shear thickening. How does this confusing behavior function in the joints where it is secreted? One interpretation is that at low movement speeds, the fluid is highly viscous to support loads, whereas at higher speeds, lower viscosity reduces the drag between the bearing surfaces while the more rapid motion helps to distribute the load (Ogston and Stanier, 1953; Barnett, 1958). During impacts, or at very high static loading, the immense shear from attempting to squeeze a very thin layer of fluid out from between the bearing surfaces could drive the fluid into its elastic state, preventing the bearing surfaces from coming into direct contact (Vogel, 2013, p. 361). Not all researchers agree with this interpretation, however. Ogston and Stanier (1953), for instance, described shear thinning as the functionally important property and the shear thickening as a biologically unimportant by-product of hyaluronic acid's molecular structure. In spite of the fairly detailed description of the mechanical behavior of synovial fluids now available, researchers are still debating the role that synovial fluid actually plays in joint operation (Vincent, 1990, p. 88; Ennos, 2012, p. 174).

4.2.4 Biological Liquids in General

Many if not most biological liquids contain at least some long chain polymers, so to the extent that they contain such polymers, they will be at least slightly non-Newtonian and shear thinning. Saliva, for example, is significantly non-Newtonian (Preetha and Banerjee, 2005). Whether saliva's non-Newtonian behavior is functionally significant, or simply a by-product of the presence of glycoproteins and enzymes needed for other functions, remains to be seen. Nevertheless, students should be aware that non-Newtonian viscosity may be more the rule than the exception for biological liquids.

4.3 MUCUS

A huge variety of animals, plants, and microorganisms secrete mucus. Mucus is a very dilute solution (approximately 2% or 3%) of glycoproteins in water. Glycoproteins, also called mucins, are proteins bound to sugars or other carbohydrates. Just as an enormous variety of organisms secrete mucus, mucus can have an enormous variety of functions. In humans alone, it is used to moisten and protect linings of air passages, trap particles in the respiratory system, and lubricate the walls and dilute food particles in the digestive system. Hagfish (eel-like relatives of lampreys) secrete large amounts of glycoproteins that can turn 10 or 20 L of water into thick mucus as an antipredator defense. Mucus forms the protective coating for the seeds of some aquatic plants and the eggs of some aquatic animals. Animals as varied as pelagic tunicates, burrowing marine annelid and echiurid worms, and sessile gastropods build particle-trapping nets from mucus. Many other suspension feeders—e.g., jellyfish, clams, sessile tunicates, sea anemones—use a layer of mucus on their particle-capturing structures to trap particles. The mucus that fish secrete over their bodies serves many protective functions and may even reduce drag when they swim (Daniel, 1981; Shephard, 1994).

4.3.1 Snail Pedal Mucus

Gastropod mollusks—snails and slugs—use mucus for a variety of functions, including adhesion, predator defense, and reproduction. Their most biomechanically interesting use of mucus is for their unique gliding form of locomotion. A snail or slug's ventral (lower) surface consists of a broad, flat, muscular surface on which it glides, called the "foot"[d]; the foot's lower surface is covered with a thin layer of mucus, the *pedal mucus*. The gliding ability of gastropods depends almost entirely on the unusual mechanical properties of the pedal mucus.

Pedal mucus can act as either a solid or a liquid depending on the circumstances. It falls in a category called "Bingham plastics" by materials scientists: it is solid up to a particular yield strain, then it suddenly yields and becomes a viscous liquid. In the case of pedal mucus, as a "solid," it is very soft, compliant, and rubbery, with a yield strength on the order of 0.001 MPa, a shear modulus of only 100−300 Pa, and a yield strain of 5 or 6 (Denny and Gosline, 1980). The actual yield point depends on the strain rate. At high strain rates, the yield strength can be more than twice as high as at low strain rates, i.e., at high strain rates it acts stronger than at lower strain rates. After the mucus yields and begins to flow, it will quickly "heal" (solidify) once the strain rate drops to zero. Within a fraction of a second it regains most of its elastic strength, but its yield strength may continue slowly increasing over tens of seconds.

When pedal mucus yields, it flows as a viscous liquid, with viscosity 3000 to 5000 times higher than water (Denny and Gosline, 1980). Not surprisingly, when flowing, the pedal mucus is non-Newtonian. It is shear thinning, with resistance to flow decreasing as flow speed (strain rate) increases. Denny

---

d. This foot inspired the name of the snail−slug taxonomic group, "Gastropoda," which can be translated as "belly-foot."

(1984) gives an equation relating experimentally measured flow stress (σ_f, closely related to shear stress) to shear rate:

$$\sigma_f = 94.9\dot{\gamma}^{0.440} \tag{4.6}$$

where $\dot{\gamma}$ is the shear rate. The exponent is less than 1.0, indicating that flow stress increases more slowly than strain rate increases, i.e., the mucus is shear thinning.

4.3.2 How Slugs Glide

The gliding locomotion of snails and slugs long mystified scientists and nonscientists alike. Key parts of the mystery were solved by Mark Denny's research in the late 1970s. Denny used enormous Northwest Pacific banana slugs, *Ariolimax columbianus*, to show that the shearing properties of pedal mucus are key to gastropod locomotion. Denny (1981, 1984) found that the bottom of a snail or slug's foot has a series of tiny waves traveling from back to front (Fig. 4.9). Each of these waves moves a tiny segment of the foot forward, while the spaces between the waves, the "interwaves," remain stationary with respect to the substrate surface. Any one point on the foot alternates between moving—as a wave passes—and being stationary. When a wave moves forward, the mucus under the wave initially and briefly strains—50–100 μm (Denny, 1981)—then yields and becomes liquid, allowing the wave to slide forward. Meanwhile, the mucus under the interwaves remains solid, and it must resist the backward push on the interwave as the adjacent wave slides forward. Thus, the exact yield stress and strain of the mucus determine how fast the wave must slide forward and how hard the interwave can push back on the mucus. In the banana slug, waves move

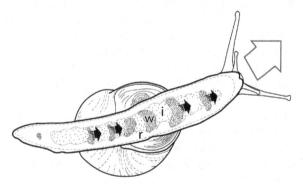

FIGURE 4.9 Ventral view of a snail's foot (functionally identical to that of a slug). The snail is gliding forward from left to right. Waves, *w* (shaded), move forward faster than the snail's overall forward speed while the rim, *r*, moves forward at the same speed as the snail's body speed. Interwaves, *i* (lightly stippled), are stationary with respect to substrate surface. *Artist: Sara Taliaferro.*

forward 3.3 times faster than the slug's overall gliding speed. Denny used empirically determined properties of slug mucus to calculate a maximum speed of slug crawling and obtained a value of 0.6 mm s^{-1}, compared with actual measured values of $0.8-2.3 \text{ mm s}^{-1}$. He commented that a relatively small change in some of his empirical coefficients could give a maximum speed of 2.3 mm s^{-1} (Denny, 1984). The mechanical properties of pedal mucus not only allow snails and slugs to use their peculiar form of locomotion but may be the major factor limiting their speed. As Denny (1984) said, "...this calculation certainly provides an argument as to why snails do not rival gazelles and cheetahs for terrestrial speed records." Recent research has improved our understanding of the details of gastropod gliding (Lai et al., 2010), but the basic pattern is still based on that described by Denny (1981, 1984).

Gastropods actually use their pedal mucus for a few functions other than locomotion. For example, some snails use mucus to seal their shells to rocks to avoid desiccation, and banana slugs form a mucus "rope" from which a pair hangs suspended while mating!

4.4 SWIMMING IN SAND: LOCOMOTION IN GRANULAR MEDIA

Dry sand can act like either a solid or a liquid: sand grains stacked up in a pile act like a solid, but if the surface on which the sand pile sits is tilted, the sand will flow like a liquid. Even though sand flowing down an incline looks very much like a liquid, the resistance of sand to an object moving through it is very different from that of a liquid. Sand's resistance is based on dry friction, not viscosity. So unlike a true liquid, the resistance of sand to an object moving through it does not change noticeably with speed, at least at biologically relevant speeds, and also unlike a liquid, resistance increases with depth of burial in sand (Albert et al., 1999). Another difference between a liquid and a granular medium of essentially spherical particles is that such particles can exist at rest (in "solid" form) at a range of different densities. In experiments using sand grain—sized glass beads, by a combination of blowing air through the container and vibrating it, researchers produces beds with volume fractions—fraction of the bed volume occupied by beads—ranging from 0.58 to 0.62 (Zhang and Goldman, 2014). (Dry quicksand, long thought to be a myth, has now been created in the lab by carefully blowing air through very fine sand to achieve volume fractions as low as 41% (Lohse et al., 2004); whether it actually exists in nature has yet to be determined.)

Many organisms move on or through granular media such as sand. For instance, the sandfish lizard, *Scincus scincus*, can both walk on the surface and burrow rapidly through dry sand. Scientists recently showed that this lizard moves through sand with lateral body undulations as if swimming through the sand, rather than digging through it with its limbs (Maladen et al., 2009). In

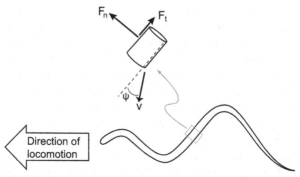

FIGURE 4.10   Resistive force theory analysis. An elongate, undulating body is divided up into many small, cylindrical segments, each with its own velocity, v, normal force, F_n, tangential force, F_t, and orientation angle, ψ. Forces and speeds are then summed over the whole body and over many time intervals to give net values for the animal. Inset: one segment at one instant.

attempting to analyze sandfish lizard locomotion through sand, Daniel I. Goldman and his students and colleagues discovered they could model the sandfish lizards' movements in sand using resistive force theory (RFT). This approach was originally devised to analyze very low Reynolds number undulatory swimming (e.g., flagellar locomotion by spermatozoa and bacteria) (Gray and Hancock, 1955), for which it achieved only mixed results (Rodenborn et al., 2013). In contrast, researchers have found that RFT does a good job of modeling movements in granular media. The RFT approach is based on dividing the undulating structure (the whole body in this case) into many small, cylindrical segments, analyzing the forces on each segment, and summing the results across all segments.[e] RFT assumes that the forces on each segment are independent of processes on all other segments, which seems a stretch for true fluids but appears to be reasonably accurate in granular media at moderate speeds. The RFT analysis represents the forces on each segment as a force normal to the segment axis (F_n) and a tangential force parallel to the segment axis (F_t), as in Fig. 4.10. Thrust comes from the component of F_n parallel to the direction of locomotion, which will be opposed by the component of F_t parallel to locomotion, representing resistance or drag. These forces depend only on the length of the segment, the segment velocity v, and the segment orientation angle, ψ (Fig. 4.10). Thus, the net force on an undulatory sand swimmer is given by

$$F = \int (dF_n + dF_t). \tag{4.7}$$

---

e. This approach is conceptually almost the same as "blade element analysis," a method of analyzing forces on helicopter rotors that has also been widely used to model forces on flapping animal wings (e.g., Sane and Dickinson, 2002).

Without constitutive equations—the functional relationships among force, speed, and orientation—to calculate the forces in sand, researchers must measure them experimentally (Maladen et al., 2009). These researchers discovered that the tangential resistance in sand was about the same as what might be seen in a viscous liquid, but the normal force was at least twice as high as in such a liquid. Applying these resistance relationships to sandfish lizard locomotion, researchers calculated undulatory sand swimming efficiency over a range of body curvatures and found that sandfish lizards use body curvatures quite close to those predicted for peak efficiency (Maladen et al., 2009); they found similar results applying RFT to a sand-swimming snake, the Mojave shovel-nose snake, *Chionactis occipitalis* (Zhang and Goldman, 2014). Moreover, when researchers built undulatory robots and tested them swimming in sand, the robots swam fastest and with highest efficiency when using the same kinematics as the sandfish lizards (Maladen et al., 2011).

The variety of types of granular media is high—sand, gravel, dust, sediment, soil, both dry and wet—and so far, studies of the mechanics of locomotion in such materials are sparse and often anecdotal. Exceptions include the work described above from Goldman's lab, as well as recent studies of animals moving in wet granular media, such as nematodes in wet sand (Jung, 2010) and polychaete worms in mud (Che and Dorgan, 2010). Given that dry sand, as Zhang and Goldman (2014) noted, can act as a solid, liquid, or gas, and that wet sand can act as a Bingham plastic (as anyone who has walked or played on a beach can confirm), the opportunities for animals in such media to exploit novel locomotion mechanics are high, and thus the opportunities for future discoveries in this field are wide.

FURTHER READING

General

Ennos, A.R., 2012. Solid Biomechanics. Princeton University Press, Princeton, New Jersey, 264 pp.

Vincent, J.F.V., 1990. Structural Biomaterials. Princeton University Press, Princeton, New Jersey, 206 pp.

Locomotion in Granular Media

Maladen, R.D., Ding, Y., Li, C., Goldman, D.I., 2009. Undulatory swimming in sand: subsurface locomotion of the sandfish lizard. Science 325, 314–318.

Non-Newtonian Fluids

Martin-Alarcon, L., Schmidt, T.A., 2016. Rheological effects of macromolecular interactions in synovial fluid. Biorheology 53, 49–67.

Secomb, T.W., 1995. Mechanics of blood flow in the microcirculation. In: Ellington, C.P., Pedley, T.J. (Eds.), Biological Fluid Dynamics. The Company of Biologists Ltd., Cambridge, UK, pp. 305–321.

Versluis, M., Blom, C., van der Meer, D., van der Weele, K., Lohse, D., 2006. Leaping shampoo and the stable Kaye effect. Journal of Statistical Mechanics: Theory and Experiment. http://dx.doi.org/10.1088/1742-5468/2006/07/P07007.

Mucus

Denny, M.W., 1984. Mechanical properties of pedal mucus and their consequences for gastropod structure and performance. American Zoologist 24, 23–36.

Lai, J.H., del Alamo, J.C., Rodriguez-Rodriguez, J., Lasheras, J.C., 2010. The mechanics of the adhesive locomotion of terrestrial gastropods. Journal of Experimental Biology 213, 3920–3933.

Shephard, K.L., 1994. Functions for fish mucus. Reviews in Fish Biology and Fisheries 4, 401–429.

Viscoelastic Materials

Alexander, R.M., 1962. Visco-elastic properties of body-wall of sea anemones. Journal of Experimental Biology 39, 373–386.

Blackledge, T.A., Hayashi, C.Y., 2006. Silken toolkits: biomechanics of silk fibers spun by the orb web spider *Argiope argentata* (Fabricius 1775). Journal of Experimental Biology 209, 2452–2461.

Espino, D.M., Shepherd, D.E.T., Hukins, D.W.L., 2014. Viscoelastic properties of bovine knee joint articular cartilage: dependency on thickness and loading frequency. BMC Musculoskeletal Disorders 15, 2–23.

Placet, V., Passard, J., Perre, P., 2007. Viscoelastic properties of green wood across the grain measured by harmonic tests in the range 0–95 degrees C: hardwood vs. softwood and normal wood vs. reaction wood. Holzforschung 61, 548–557.

Chapter 5

Systems and Scaling

5.1 PUTTING IT ALL TOGETHER: BIOMECHANICS IN ACTION

Up to this point, we have looked at the biomechanical properties of materials and individual components of organisms—struts, shells, blood, mucus—more or less in isolation. In this chapter, we will look at examples of components fitting together into functional systems and see how organisms can both exploit and be constrained by biomechanics and scaling. Some researchers have suggested that, because natural selection can produce optimized systems, a thorough understanding of organismal biomechanics might lead to useful, practical products for humans. This suggestion is the basis for the relatively new field of *biomimicry*, with which we will conclude this chapter.

5.2 LEGS: MUSCLES, JOINTS, AND LOCOMOTION

Two major groups of animals have independently evolved terrestrial loco-motion using rigid, jointed limbs: arthropods and vertebrates. (Indeed, the very name "arthropod" is usually translated as "jointed limb" or "jointed foot.") Such a limb requires three components: muscles to produce the force and motion, rigid beams to form supporting segments, and joints or articulations to allow movement. Arthropods and vertebrates arrange these components rather differently—for instance, arthropods put the muscles inside the beams whereas vertebrates have them on the outside—but do not let these differences obscure the fact that both groups build their legs based on the same three fundamental components.

5.2.1 Muscle Biomechanics and Scaling

The mechanical function of muscle tissue is based on the operation of the protein filaments actin and myosin in the sliding-filament mechanism, but a description of these subcellular processes is beyond the scope of this book (such a description can be found in any modern physiology textbook, e.g., Sherwood et al. (2013)). Although entire books have been written about the mechanical properties of muscle tissue (e.g., Herzog, 2000), our more limited

Nature's Machines. http://dx.doi.org/10.1016/B978-0-12-804404-9.00005-0

121

focus will be on an overview of the scaling of muscle mechanics. To avoid confusion, biologists use "muscle tissue" to refer to an assemblage of those specialized, mechanically active cells powered by actin and myosin, and "a muscle" to refer to an organ of the skeletal muscular system—consisting mostly of muscle tissue, but also including connective tissues, blood vessels, etc.

Muscle tissue functions by shortening, or producing active tension, or a combination. Muscles can only pull, they cannot push, which is why muscles are usually arranged in antagonistic pairs; once a muscle has shortened, it requires an antagonistic muscle to reverse the movement. Sometimes an elastic component can serve as an antagonist, such as the elastic hinge ligament that is the antagonist of the shell-closing muscle of a clam, but such passive antagonists are the exception, and opposing muscles are the usual antagonists.

For a given muscle, its force (tension) and speed of shortening will be our main focus, along with the product of force times speed, i.e., power. A fundamental property of muscle tissue is that its maximum force production is proportional to its cross-sectional area, so we can express a muscle's tension as a force per unit area (equals stress). Moreover, within vertebrates, the proportionality is very nearly constant for all skeletal muscles, approximately 250–300 kPa (Goldspink, 1980; Vogel, 2001). Recall from Chapter 2 that collagen, the main component of tendons, has a tensile strength of approximately 100 MPa, which helps explain why tendons can be so much narrower than the muscles to which they attach. Invertebrates show more variation in maximum tensile stress—due to more variation in the dimensions of the sliding filament arrangement (Vogel, 2001, p. 71)—so have evolved unusually forceful but slow muscles such as crab claw closer muscles (Taylor, 2000) or very rapid but weak muscles, such as those used for rapid extension of squid tentacles (Kier, 1985). Nevertheless, the general rule still holds: the maximum force production of a particular type of muscle is proportional to the cross-sectional area, even if the constant of proportionality might be different for some invertebrate muscles.

The way the force output of muscle scales with size helps explain the apparently outsized strength of an ant. Whereas a human may struggle to lift $\frac{1}{3}$ to $\frac{1}{2}$ of his or her body weight, ants appear to carry loads of up to 10 times their body weight with ease. Does the ant have much more powerful muscles? Consider the size difference: a person will be somewhere between 1 and 2 m tall, whereas the ant might be around 2 mm (approximately 1/10 inch) long. The ant will thus be between 500 and 1000 times shorter than a person. If the ant were geometrically similar to a human, its cross-sectional area (and hence, the cross-sectional area of its muscles) would be proportional to (body length)^2, or at least 250,000 times smaller. Assuming body mass is directly proportional to body volume, its body mass would be proportional to (body length)^3 or 125 million times smaller than a human. A geometrically similar ant's muscle force would therefore be a quarter million times smaller but its weight would be over 100 million times smaller, so by this argument, such an ant ought to be able to lift 500 times its own weight. Obviously, ants are nowhere near geometrically similar to humans, and ant muscles are actually much narrower for their mass than human muscles. An ant has no need to be strong enough to lift 500 times its body weight, so the ant will get by just fine using muscles with such low cross section that, if scaled up to human size, they would be far too weak to power locomotion.

In contrast to force production, the maximum contraction speed of different muscles can vary quite a bit. To compare muscles, the intrinsic speed of shortening, v_{IS}, normalizes speed by dividing the maximum shortening speed by the muscle's resting length, giving units of inverse seconds (s^{-1}, but usually spoken as "lengths per second"). Different muscles can have different intrinsic speeds, largely due to the rate at which the myosin of different muscles consumes adenosine triphosphate (Goldspink, 1980).

If all muscles used sliding filaments that shortened at the same rate, then a geometric argument shows that v_{IS} should be proportional to (body length)$^{\frac{1}{2}}$, meaning that smaller animals should have faster muscles (Close, 1972), which does seem to hold approximately for terrestrial mammals (Hill, 1950). Across the animal kingdom, however, enough variation in the speed of sliding filaments exists that there is no simple proportionality between v_{IS} and body size, other than the general observation that small animals tend to have faster muscles than large animals (Goldspink, 1980).

One fundamental aspect of muscle speed and force production is that muscle tissue produces its maximum force at zero speed and its maximum speed at zero force (Fig. 5.1). This contrasts sharply with internal combustion

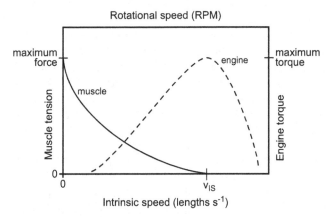

FIGURE 5.1 Force versus speed for muscle tissue compared with an internal combustion engine. *Solid curve* shows tension (force) versus intrinsic speed for muscle tissue. *Dashed curve* shows torque (force × distance) versus rotational speed for an engine. Curves have been scaled, so maximum tension or torque occurs at the same y-axis location for both curves. v_{IS}, maximum intrinsic speed.

engines, which cannot produce any force (strictly speaking, torque, for a rotating machine) at zero rotational speed and can also produce some torque at speeds well above their speed of maximum torque (Fig. 5.1). Moreover, because power is force times velocity, muscle tissue produces its maximum power at some speed between zero and its maximum speed, and for typical muscles, maximum power occurs at a shortening speed of about $\frac{1}{3}$ its maximum speed. Most animals tend to use their locomotion muscles at a shortening speed that produces maximum power.

Values for maximum power output from isolated muscle tissue can be quite misleading. The maximum specific power output (power output per unit mass) of an isolated muscle will generally be in the range of $200-500$ W kg^{-1} (Goldspink, 1980; Vogel, 2001). In an intact animal, however, that much power is only available very briefly, perhaps a second or less. For any activities lasting more than a couple of seconds, muscle power is limited by the ability of the body to supply oxygen and fuel. Thus, actual sustained power outputs rarely reach even one-tenth of the theoretical maximum for activities of more than a minute or two in duration. Champion athletes might maintain 10 W kg^{-1} for an hour (Wilkie, 1960), but an average human laborer would have great difficulty exceeding 3 or 4 W kg^{-1} over a full day of work (Vogel, 2013, p. 466). Within mammals, the maximum specific power for isolated muscles seems to be fairly constant, but the greater variation in v_{IS} and maximum stress among invertebrates again produces more variation in specific power among those groups.

5.2.2 Articulations: Adding Flexibility to Rigid Skeletons

Appendages supported by rigid skeletons require joints or *articulations* to allow adequate movement. Although arthropods and terrestrial vertebrates have both evolved articulations so they can use their legs for walking, the structural arrangements of their articulations are very different.

Arthropods are either small or aquatic, so the gravitational loads on their skeletons are quite low relative to those on vertebrates. As mentioned in Chapter 2, small arthropods sometimes construct joints in their otherwise cylindrical legs by forming a short region with an oval or flattened cross section and slightly more compliant cuticle. The animal bends the joint using controlled local buckling, with no need for pivots or sockets. Such joints only work for small loads and dimensions and are generally limited to small aquatic arthropods (Wainwright et al., 1982, p. 278).

Most arthropod legs make use of a series of *dicondylic* joints. These consist of a small peg and socket (the "condyles") on each side of the joint, straddling regions of arthrodial (flexible) cuticle (Fig. 5.2). Such a joint acts as a hinge with a single degree of freedom (Currey, 1980a), i.e., it can only bend in one plane. For a limb to be able to bend in more than one plane, e.g., front—back and up—down, the typical insect or crab leg requires an additional joint with its

(A)   **(B)**

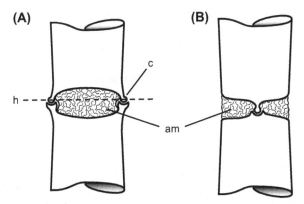

FIGURE 5.2 Arthropod dicondylic joint. The exoskeleton of leg segments above and below the joint, including the condyles, is made of tanned (rigid) cuticle; the exoskeleton of the joint region between condyles, the arthrodial membrane, is made of untanned, compliant (flexible) cuticle. (A) Front view. (B) Side view. *am*, arthrodial membrane; *c*, condyle; *h*, hinge axis.

hinge axis perpendicular to that of the first joint. Very few arthropod limb joints operate with more than a single degree of freedom, but many insect legs have a pair of dicondylic joints close together where they function very much like the engineer's universal joint (Ennos, 2012, p. 174).

Insects can get by with joints that use tiny pegs and sockets because their joint loads are small, they can make cuticle very hard, and the bearing surfaces are so close to the axis of rotation that frictional forces are miniscule. Terrestrial vertebrates, in contrast, use a much weaker material—cartilage—to cover bone ends at joints. This material difference requires a different structure. The weaker cartilage and larger gravitational loads requires the load to be spread out over a larger surface area. Large joint surface area requires a large radius of curvature for the bearing surface for joints that flex or rotate (Currey, 1980a). Thus, vertebrate limb joints incorporate knobby enlargements at the end of the bone. This gives the bearing surfaces much larger surface area than the cross section of the bone shaft because the cartilage of the joint cannot withstand as high of stresses as the bone. Moreover, to prevent the cartilage from wearing away, such joints need good lubrication, which is provided by the synovial fluid. Although the exact mechanism of joint lubrication is still being worked out, the net result is a very low-friction structure, with friction coefficients (ratio of friction force to applied perpendicular load) of 0.1 or less (Ennos, 2012, p. 174).

Vertebrates have evolved a much wider diversity of joint types than arthropods. Ball and socket joints, quite rare in arthropods, are common among vertebrates; our hip and shoulder joints are examples. Such a joint has three degrees of freedom: bending in two perpendicular planes—up–down and left–right—plus rotation. At the other extreme are simple hinge joints with one degree of freedom, such as the joints in our fingers. Superficially, the human knee appears to be a simple hinge joint, but it actually allows significant rotation when flexed to 90 degrees.

Arthropod joints are embedded in the exoskeleton, which holds them together. Vertebrate joints, on the other hand, have a set of cord- or straplike ligaments that both hold the bones together at the joint and place limits on joint movement. (Ligaments, like tendons, are mostly made of collagen, and the difference is largely semantic: ligaments connect bones to bones, tendons connect muscles to bones.) Aside from some ball and socket joints where the ball is captured by the lip of the socket, most vertebrate joints will fall apart if all the ligaments are removed. Many ligaments do more than simply hold the bones close to each other. By placing attachments in different locations relative to the joint's axis of rotation, a ligament can become taut—limiting further travel—in one direction of movement or the other, or both (Alexander and Bennett, 1987).

5.2.3 Locomotion on Two (or More) Legs

Legged locomotion always involves some variation on placing a foot on the ground and pushing forward, then lifting it up, swinging it forward, and placing it on the ground again. Humans do it with two legs (so are thus bipedal), horses do it with four legs (quadrupedal), and cockroaches do it with six legs (hexapedal). The *stance* or *support* phase is when a foot is on the ground, and the *swing* or *recovery* phase is when a foot is off the ground and swinging forward. A *stride* is a complete stepping cycle for one leg, from the end of one support phase through the swing phase to the end of the next support phase. *Stride length* is thus the distance between where a foot touches the ground during one stance phase and where it touches the ground in the next stance phase. The *duty factor* is the fraction of a stride cycle where the foot is on the ground: a duty factor of 0.6 means the foot is on the ground for 60% of the stride cycle, whereas a duty factor of 0.3 means the support phase only lasts 30% of the stride cycle.

We begin with bipedal locomotion, both because it is what humans do, and it has fewer permutations than locomotion with more legs. Bipeds, such as humans or ostriches, can use one of two *gaits* or stepping patterns. A *walking gait* is defined as a stepping pattern with a duty factor of >0.5, meaning that at least one foot is on the ground at all times, and at least some part of the stride includes a double stance phase where both feet are on the ground simultaneously. A *running gait* is a stepping pattern with a duty factor of <0.5, which means the stride includes a period called the *aerial* phase when neither foot is on the ground.[a]

---

a. Although people refer to "jogging" or "trotting" gaits in humans, mechanically those are just slow runs and not separate gaits, in contrast to quadrupedal trotting, which is actually a separate gait, one of several possible quadrupedal running gaits.

The walking and running gaits differ dramatically in how the body's center of mass moves, which translates into very different relationships between kinetic and potential energy changes. Walking is often described as an inverted-pendulum gait, also called a compass gate because in its simplest form, the hip follows a circular path as if drawn by a drafting compass. When a pendulum swings back and forth due to gravity, its potential energy is highest at the top of its swing, just where the kinetic energy passes through zero. At the bottom of its swing, the potential energy is at its minimum but its speed is at its highest so kinetic energy is at its peak. Now consider the stiff-legged walk of Fig. 5.3A. When the front foot hits the ground, the body's center of mass and hence its potential energy will be at their lowest (Fig. 5.3B), but

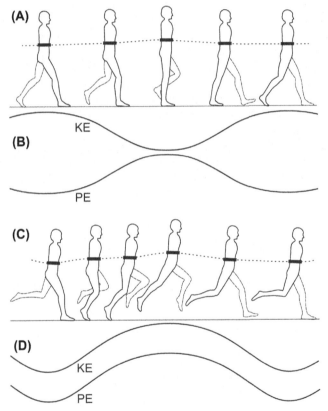

FIGURE 5.3 Kinematics and energetics of bipedal walking and running gaits. (A) Kinematics of inverted pendulum walking. *Dark horizontal bar* represents body's center of mass. (B) Corresponding changes in kinetic energy (KE) and potential energy (PE) during walking; KE and PE are out of phase. (C) Kinematics of bipedal running. (D) Corresponding KE and PE change during running; KE and PE are in phase. *Artist: Sara Taliaferro.*

speed, and hence kinetic energy, will be at their highest because the center of mass has been accelerated by gravity. This kinetic energy pushes the center of mass forward and up, using the leg like a pole-vaulter's pole. Gravity then decelerates the center of mass so that at the peak of the stride, speed (and kinetic energy) is at a minimum just when center of mass height (and potential energy) is at a maximum. Past the peak, potential energy decreases as the center of mass accelerates downward, increasing kinetic energy. Thus, in the walking gait, kinetic and potential energy are constantly being traded back and forth. Studies of human walking show that energy is about 65% conserved, the missing 35% going into accelerating and decelerating the legs and body (McMahon, 1984, p. 191). In fact, without friction (in joints, with the ground, etc.) walking on the level would, in principle, require no work at all except to get the cycle started. This is nicely illustrated by a walking toy (Fig. 5.4), that, if given a slight nudge to start a sideways rocking, will walk down a ramp passively, powered entirely by gravity. Dubbed as "passive dynamic walkers" by researchers, they have been analyzed extensively by roboticists (Tedrake et al., 2004). Human walking is actually not a perfect compass gait because complex movements of the hips slightly flatten the arc to smooth out the movements and slightly increase stride length (McMahon, 1984, pp. 193−196); nevertheless, the inverted-pendulum mechanism gives a good qualitative description of the potential energy and kinetic energy cycling during human walking.

When we walk on a flat, level surface, we use muscles to absorb the deceleration when a foot contacts the ground, to add a bit of force to push off, and to start the leg swinging forward, but these inputs are relatively small and

FIGURE 5.4 Ramp-walking toy, also called a passive dynamic walker. Legs are freely hinged to the horizontal axle, which serves functionally as the hip joint. It can walk down an incline without any motor or actuator. *Artist: Sara Taliaferro.*

often isometric. Gravity dominates the motion, decelerating the center of mass as it moves up and accelerating it as it falls back down. In fact, gravity sets the maximum speed of an inverted pendulum walk. If an animal's leg length is l and it walks at speed v, the body's center of mass moves in an arc of radius l, which requires a centripetal acceleration toward the center of the arc of v^2/l. With gravity providing the only downward force, the highest acceleration possible is the acceleration of gravity, so if $g = v^2/l$, then the maximum speed will be

$$v_{max} = \sqrt{gl}, \tag{5.1}$$

which for humans works out to about 3 m s^{-1} (roughly $6\frac{1}{2}$ mph) (Alexander, 1982, p. 88). If a walker tries to walk faster than this maximum speed, he or she will tend to bounce off the ground instead of keeping the foot firmly planted. (Walking racers use exaggerated hip movements to minimize vertical changes in center of mass and effectively increase l, which allows faster walking but at very high metabolic cost.) The easiest way to move faster is to switch to a gait that is not powered by gravity, i.e., a running gait.

In the running gait, the runner pushes off at the end of the stance phase, lifting the body off the ground into the aerial phase and elevating the center of mass (Fig. 5.3C). As the body falls back down at the end of the aerial phase, the stance leg hits the ground and bends to decelerate the body. Then the stance leg straightens to push off and begins the next aerial phase. A bipedal runner thus has two aerial phases per stride. In a run, the kinetic energy and potential energy are in phase (Fig. 5.3D): at the end of the aerial phase, the falling body causes the stance leg to bend and decelerate the body, so both kinetic and potential energy will be at a minimum. When the leg straightens out during the push off, the body's center of mass accelerates up and forward, so both kinetic and potential energy will peak at the top of the aerial phase. Because the kinetic and potential energy are in phase, the total energy of the body's center of mass goes through large changes, and potential and kinetic energy cannot be traded off against each other as they can in a walk. If all the work of decelerating and accelerating the body were done by muscles, running might be a very metabolically expensive process, but most running animals have a different way to conserve energy: using springs.

Imagine a runner's legs replaced with passive compression springs. The spring could be compressed at the beginning of the stance phase when the runner lands, storing energy. Then the spring could extend and release the energy for the push off into the next aerial phase. In fact, extensive research has shown that many runners—ranging from kangaroo rats to ostriches and horses—do store and release energy elastically during running (Cavagna et al., 1977; Alexander et al., 1979; Biewener et al., 1981; Dimery et al., 1986; Ker et al., 1987). In humans, as in most other running mammals, the Achilles tendon on the back of the lower leg acts as the main spring, with some assistance from other tendons and ligaments. In the neighborhood of 40%

—50% of the energy needed for running is stored and released passively, thus reducing the amount of work the muscles need to contribute by that amount. For example, kangaroos use a hopping gait where both legs swing and support in phase (rather than in alternation as in human running), but the processes that go on in each leg are essentially the same. Biewener et al. (1998b) showed that elastic energy storage in wallabies (small kangaroo relatives) saved them at least 45% of the total work of hopping, and running humans can obtain up to 52% of the work of running by storing energy in the Achilles tendon and the ligaments of the arch of the foot (Ker et al., 1987).

Four-legged or quadrupedal animals have many possible stepping patterns (see Box 5.1). Most use three basic gaits: walking, trotting, and galloping. A quadrupedal walk involves moving one foot at a time, so the animal always has at least three feet on the ground at any time. Starting with moving the left front leg, the normal stepping pattern is left front, right hind, right front, left hind. Duty factors are always more than 0.5, such as 0.7 for a slow walking dog (Alexander and Jayes, 1983). Quadrupeds typically use two running gaits, the

---

BOX 5.1 Other Gaits

Stepping patterns besides the bipedal walk and run, and the quadrupedal walk, trot, and run, are possible and some are used by particular animals. Bipedal hopping—swinging and planting both legs simultaneously—is used by kangaroos and relatives and by several small rodent species. Hopping is energetically the same as running, but usually with greater rise and fall of the body center of mass and with longer stride lengths.

Among quadrupeds, many more combinations are possible. The *rack*, also called the pace, is like a trot except that the front and hind legs on the same side, rather than opposite sides, move together. Camels and giraffes rack instead of trotting, apparently because their long front and hind legs might interfere with each other and limit stride length in trotting. Other quadrupeds occasionally rack naturally; horses can be trained to rack (McMahon, 1984), but the side-to-side motion makes for a distinctly uncomfortable ride, so it is not seen much outside of harness racing.

In the *canter*, one diagonal leg pair moves together in phase, while the other fore and hind legs each move separately, e.g., the right hind leg alone, then the left hind and right fore together, then the left fore alone. Although the footfall pattern seems intermediate between trotting and galloping, dynamically it is sometimes considered a form of gallop.

At their top speeds, small mammals such as mice and rabbits use another gait similar to galloping, the *bounding* gait. In a bound, the forelegs move exactly in phase and the hind legs move exactly in phase. Why bounding is limited to small mammals and galloping, to medium and large animals, is not entirely clear, but may reflect scaling of energetics, foot loading, or some combination (McGreer, 1992).

trot at intermediate speeds and the gallop at high speeds. The trot is called a symmetrical gait because the right and left legs of each pair move exactly one half stride out of phase, and each moves with its diagonally opposite partner (Alexander, 2003, p. 110): the left front and right hind swing together, then the right front and left hind swing together. Each diagonal pair of legs has a duty factor of less than 0.5, so a trotting animal has two aerial phases per stride. In a gallop, the animal switches from swinging a fore leg and hind leg together to swinging both front legs approximately together and both hind legs approximately together. When medium-sized and large quadrupeds such as dogs and horses gallop, the front pair and hind pair of legs are not exactly synchronized: one front foot lands slightly before the other and one hind foot lands slightly before the other. Duty factors are lower in galloping than trotting (as low as 0.2 or 0.3 (Alexander and Jayes, 1983)) and because of the near synchrony of the front pair of legs and hind pair of legs, galloping animals can flex and extend their spinal columns and rotate their hip and shoulder girdles to increase their stride lengths (and hence, speed). Cheetahs use this to great effect, gaining up to 20% greater distance per stride (Biewener, 2003, p. 57). Many gallopers also use their spines as another spring for storing elastic energy, flexing it when the front legs are on the ground and extending it when the hind legs are pushing off. (Larger animals such as horses need a stiff spine for weight support and generally do not flex their spines much during galloping, fortunately for the humans that ride horses.)

We saw earlier that gravity places an upper limit on walking speed, with $v_{max} = (gl)^{1/2}$. This can be expressed in terms of another dimensionless index, the Froude number,[b] Fr:

$$Fr = \frac{v^2}{gl} \qquad (5.2)$$

where l in this case is leg length. Thus, the maximum walking speed occurs at $Fr = 1$. In fact, most bipeds switch to running, and quadrupeds switch to trotting, at Froude numbers of about 0.5 or 0.6 (Alexander, 2003, p. 109). The transition from one gait to another appears to depend mainly on the relationship between metabolic costs and speed: as Fig. 5.5 shows, at low speeds, walking is cheaper, but above the intersection of the walk and run curves, running is cheaper. This transition appears to occur at a similar Froude number for both bipeds and quadrupeds. Similarly, horses switch from trotting to galloping at speeds where galloping becomes metabolically cheaper (Hoyt and Taylor, 1981). The transition from trot to gallop seems to occur at Froude

---

b. The Froude number was originally developed to analyze ship speeds, which are limited by the effects of gravity on bow wave and wake formation. Researchers have since discovered that it can be applied to many other situations where speed and gravity interact. In some fields, the square root of the right side of Eq. (5.2) is used, but there does not seem to be a convention for the preferred form in biomechanics.

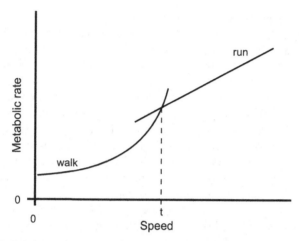

FIGURE 5.5 Relationship between metabolic rate and speed for different bipedal gaits in humans. Below the transition speed (t), walking is metabolically cheaper, and at speeds above the transition, running is metabolically cheaper; t corresponds to a Froude number of approximately 0.5.

numbers near 2.5, although with a bit more scatter than the walk–run transition Froude number.

Many arthropods have more than four legs: insect have six, spiders have eight. Insects are well known for using a "tripod" gait, where three legs at a time move while the other three remain on the ground. The support legs always consist of the front and hind legs on one side and the middle leg on the opposite side. The support legs thus form a tripod, which, like a three-legged stool, is very stable. As long as the insect's center of mass is inside the tripod formed by the legs, the insect cannot fall over—it is statically stable and does not need to expend any energy to keep its balance. The stepping pattern is thus left fore, left hind, and right middle legs all swinging together, then right fore, right hind, and left middle legs swinging together. Insects can use other gaits, but the tripod gait seem to be by far the most widely used, and many insects use it for a wide range of locomotion speeds. For an interesting exception, including bipedal cockroaches, see Full and Tu (1991).

5.3 "SOFT" (HYDROSTATIC) SKELETONS

Humans have a rigid internal skeleton of bone, so we are used to thinking of columns—like our legs—or cantilevers—like our arms—supported by hard, internal structures. Not all animals follow this pattern. Many support their bodies using pliant or flexible materials. The mesoglea of a sea anemone described in the previous chapter is a simple example. Another common example is the *hydrostatic skeleton*, capable of much greater force transmission and faster and more complex movements than those supported by mesoglea.

5.3.1 Muscles and Stresses in the Wall

Hydrostatic skeletons (sometimes just called "hydrostats") use a cavity filled with water; the water is incompressible, so the organism can use it to apply force or change shape. Plants use osmotic pressure to pressurize the cavity, whereas animals do it with muscle layers in the hydrostat's walls. The most common muscle arrangement is to have a layer with lengthwise or longitudinal fibers and a layer with circular or circumferential fibers. Most hydrostatic skeletons are more or less cylindrical, so longitudinal muscles will tend to shorten them and also widen them due to constant volume, whereas circumferential muscles will tend to do the opposite. The longitudinal and circumferential muscles are thus antagonistic.

When muscles surrounding a cylindrical, water-filled cavity contract, they can increase the pressure in the cavity. The stresses in the wall of a pressurized cylinder are given by the Laplace equations, which arise from basic geometry (Ennos, 2012, p. 104). The circumferential stress, σ_C, is given by

$$\sigma_C = \frac{PR}{t} \tag{5.3}$$

where P is the pressure inside the cylinder, R is the radius of the cylinder, and t is the thickness of the cylinder's wall. The longitudinal stress, σ_L, is given by

$$\sigma_L = \frac{PR}{2t} \tag{5.4}$$

so the longitudinal stress in a cylinder's wall is half the circumferential stress. This difference can lead to aneurisms; when the circumferential stress reaches the tensile strength of the wall, a spherical bulge, the *aneurism*, appears. This may occur when inflating a cylindrical balloon. At first, the pressure just straightens the balloon, but as you blow harder and increase the pressure, a spherical enlargement suddenly appears. The stress in the wall of a sphere is the same as the longitudinal stress in a cylinder, so an aneurism initially reduces the circumferential stress when it starts to form. Because it will also decrease the wall thickness, however, it can still lead to tensile failure and wall rupture, depending on the material. (A rubber balloon does not rupture because of rubber's sharp rise in stiffness at large strains; less extensible materials are at greater risk of failure.)

5.3.2 Fiber-Reinforced Hydrostats

If the walls of a hydrostat are reinforced with tension-resisting fibers, then pressurizing the hydrostat can produce a stiff structure. Organisms can thus build a stiff body or part without the need for metabolically expensive material such as bone or shell, using only "cheap" tensile fibers. All plant and most animal hydrostats have walls reinforced with tensile fibers—cellulose in plants, usually collagen in animals. One simple way to arrange such fibers is

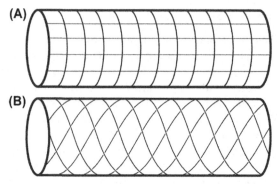

FIGURE 5.6 Possible arrangements of reinforcing fibers around a cylinder. (A) The orthogonal pattern, with longitudinal fibers at right angles to circumferential fibers. (B) A crossed-fiber helical array of reinforcing fibers, with some of the fibers forming a right-hand helix and other forming a left-hand helix. *Artist: Sara Taliaferro.*

to have hoop-shaped fibers running around the circumference to resist expansion in diameter and longitudinal fibers running lengthwise to resist lengthening; this is the "orthogonal" reinforcing arrangement (Fig. 5.6). The hoop fibers resist increase in diameter and the longitudinal fibers resist increase in length, but they do not resist torsion, i.e., twisting movements. Such a hydrostat is quite resistant to compression, extension, and bending but is prone to local buckling or kinking, as well as twisting. This arrangement is actually very rare in nature, so far only found in mammalian penises (Kelly, 2007).

A much more common arrangement has the fibers wrapped helically around the cylinder corkscrew fashion, with some fibers in left-handed helices and others in right-handed helices, a so-called *crossed fiber helical array* (Kier, 2012), as shown in Fig. 5.6B. Almost all biological hydrostats (aside from muscular hydrostats, which we will discuss separately) use such crossed-fiber reinforcing arrays. To function properly, the fibers must have at least some freedom to slide past each other in the matrix, and they must be essentially inextensible at biological pressures. Helically wound hydrostats are the complete converse of orthogonally reinforced hydrostats: helically wound hydrostats resist twisting (because of the angled fiber orientation) and kinking, form smooth curves under bending loads, and under some circumstances can change length and diameter even though the fibers themselves are inextensible.

The fiber angle of a helical reinforcing array plays a critical role in the hydrostat's response to pressure changes. The fiber angle is the angle the fiber makes with the long axis of the cylinder and is normally the same for both the left-turning and right-turning helices of a given hydrostat (Fig. 5.7). Geometry dictates that the volume of the cylinder will be maximized when the angle of the reinforcing fibers is 54.7 degrees relative to the cylinder's long axis. This is

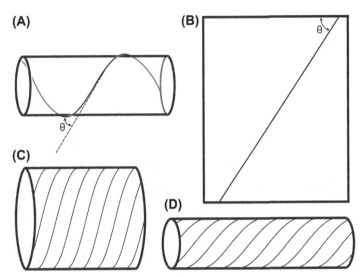

FIGURE 5.7   Helical fiber angles. (A) A single helical fiber, with a fiber angle, θ, of approximately 55 degrees to the cylinder's long axis. (B) The surface of the cylinder in (A) unrolled, showing the fiber and fiber angle, θ. (C) A cylinder with a high fiber angle ($\theta > 55$ degrees) will tend to get longer and narrower as pressure increases. (D) A cylinder with a low fiber angle ($\theta < 55$ degrees) will tend to get shorter and wider as pressure increases. (In (C) and (D), for clarity, only fibers with helices in the same direction are shown.) *Artist: Sara Taliaferro.*

a strictly geometric constraint that has nothing to do with material properties, as long as the fibers are basically inextensible. Because maximum volume occurs at a fiber angle of approximately 55 degrees, if the fiber angle is high—lots of turns in a given length—then increasing the pressure will tend to lengthen and narrow the cylinder to make the fiber angle closer to 55 degrees (Fig. 5.7). If the fiber angle starts out lower than 55 degrees, increasing the pressure will tend to shorten and widen the cylinder. This relationship is shown quantitatively in Fig. 5.8 (Fig. 5.8A is the standard graph, and Fig. 5.8B shows the same data replotted in what Vogel (2003) considers to be a more intuitive form). Hydrostats in the flaccid region—under the curve—have an oval or noncircular cross section, whereas hydrostats on the curve are turgid or fully inflated, with a circular cross section. If a hydrostat is in the flaccid region under the curve, increasing pressure tends to first inflate the hydrostat until it reaches the curve and achieves a circular cross section and then moves it up the curve toward the peak. (It cannot go outside the curve because that would require it to be more than fully inflated, i.e., ruptured.) If its fiber angle starts out less than 55 degrees when it becomes turgid (i.e., when it reaches the curve), moving up the curve will cause it to become shorter and wider, but if its fiber angle is greater than 55 degrees when it reaches the curve, it will lengthen and become narrower as it moves up the other side of the curve.

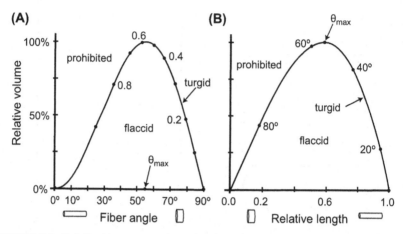

FIGURE 5.8 Relationships among volume, length, and fiber angle for a helically reinforced hydrostat. (A) Standard graph with volume as a function of fiber angle, and relative length shown on the curve parametrically. Cylinders get longer and narrower to the left, and shorter and wider to the right. (B) The same data, replotted with relative volume as a function of relative length, and fiber angle shown parametrically; cylinders get shorter and wider to the left, and longer and narrower to the right (Vogel (2003) considers this version to be more intuitive). A relative length of 1.0 is the length of exactly one turn of a reinforcing fiber around the circumference; at a fiber angle of 0 degree, the cylinder has a volume of zero and becomes a line the length of the fiber, whereas at a fiber angle of 90 degrees, the cylinder becomes a disk of zero volume and circumference of the length of the fiber. θ_{max}, fiber angle that gives maximum volume, 54.7 degrees. *Redrawn from data in Clark, R.B., Cowey, J.B., 1958. Factors controlling the change of shape of certain nemertean and turbellarian worms. Journal of Experimental Biology 35, 731–748, Wainwright, S.A., Biggs, W.D., Curry, J.D., Gosline, J.M., 1982. Mechanical Design in Organisms (Paperback Ed.). Princeton University Press, Princeton, New Jersey, 423 pp., Vogel, S., 1998. Cats' Paws and Catapults: Mechanical Worlds of Nature and People. W.W. Norton & Co., New York, 382 pp., and Vogel, S., 2003. Comparative Biomechanics: Life's Physical World. Princeton University Press, Princeton, New Jersey, 580 pp.*

5.3.3 Biological Examples

Oddly enough, biological hydrostats mostly use fiber angles well away from 55 degrees. Nematodes (roundworms), for example, have a fiber-reinforced outer body covering or cuticle with a fiber angle of approximately 75 degrees. Their only muscles are longitudinal, which would tend to shorten the body if they all contracted together. Because of the high fiber angle, the body would need to increase in volume to shorten, but with body contents mostly of water, its volume cannot change. So when the muscles contract, they do not change the body's length, but do produce a large increase in pressure (up to 30 kPa), maintained by continuous partial contraction of the muscles. The high pressure stiffens the body and opposes bending. Then, when muscles on one side of the body increase their contraction, the body will form a smooth curve with the fiber-reinforced cuticle acting as an antagonist to the longitudinal muscles. In life, nematodes squirm though their environments using waves of bending, with only longitudinal muscles and without getting shorter (Ennos, 2012, pp. 119–120).

In contrast, the outer body wall or "mantle" of a squid contains a helical array of collagen fibers with a fiber angle at rest of approximately 27 degrees and a layer of circumferential muscle. To expel water from the mantle cavity for swimming, the squid contracts the circumferential muscles. This increases the pressure, which would lengthen the mantle cavity in the absence of fibers. The fibers prevent the cavity from lengthening, however, so water is expelled (forcefully) through a narrow aperture to decrease the volume (Ward and Wainwright, 1972), in the process forming a jet for locomotion.

Some worms, such as nematodes and earthworms, maintain a more-or-less cylindrical body, but others, such as flatworms (Platyhelminthes) and ribbon worms (Nemertea), have a flattened shape at rest. These flattened worms have helical reinforcing fibers with fiber angles close to 55 degrees when relaxed, but their noncylindrical shape means that they are in the flaccid region of Fig. 5.8. They have layers of both longitudinal and circumferential muscle, with the curious property that when either contracts, it makes the worm more circular in cross section. If the longitudinal muscle contracts, the worm gets shorter, moving it on a horizontal line (to the right on Fig. 5.8A or the left on Fig. 5.8B) toward the curve, making it both more circular and getting stiffer as it approaches the curve. If the circumferential muscles contract, the body gets longer and it moves horizontally (to the left on Fig. 5.8A or the right on Fig. 5.8B), this time getting longer and rounder, and again getting stiffer as it approaches the curve (Clark and Cowey, 1958). In life, the worms may not ever reach the curve to become fully inflated, but this relationship between shape and stiffness plays a significant role in their movements.

Researchers have cataloged a wide variety of organisms that use helically reinforced hydrostats, ranging from sunflower stems to shark bodies. For good general discussions of the topic, see Vogel (2013) or Ennos (2012). The latter discusses the role played by helical reinforcing fibers in controlling the growth form of cells in growing stems of herbaceous plants.

5.3.4 Muscular Hydrostats

Imagine a cylinder made entirely of solid muscle tissue, with cells arranged in sheets or bundles running in various directions. The muscle tissue, being mostly water, is for all practical purposes incompressible; the whole structure can function as a hydrostat. This type of structure can function simultaneously as a structural element and as a source of force and movement. Such structures are called *muscular hydrostats* (Kier and Smith, 1985), and they make up such varied body parts as lizard and mammal tongues, elephant trunks, squid tentacles, and octopus arms. They usually have muscle sheets or bundles running lengthwise (longitudinal), perpendicular to the long axis (either circumferential or across the diameter) and at an angle to the long axis (oblique or helical). When the longitudinal muscles all contract, it shortens; when the perpendicular muscles contract, it lengthens; and when one set of the helical muscles contracts, it twists. If longitudinal muscles on one side contract, the whole hydrostat bends. If both sets of helical muscles contract, it stiffens—and it may also lengthen or shorten, depending on the fiber angle.

We know that muscle tissue can only shorten, but perpendicular muscles in a muscular hydrostat cause the whole cylinder to elongate. This arrangement thus gives the appearance of a muscle that can actively elongate or push. Perpendicular muscles can also function as a sort of virtual lever system. Vogel (2013, pp. 419–420) described how changes in circumference affect length and force for a constant-volume cylinder. He shows that for a short, fat cylinder (length < diameter), a unit decrease in circumference produces less than a unit increase in length while increasing the force. The force increase will be inversely proportional to the distance increase, just as the force increase is inversely proportional to the lever arm for a conventional lever system. Conversely, for long, skinny hydrostats (length > diameter), a unit decrease in circumference can generate more than a unit increase in length, in this case with a proportional decrease in force. This again is analogous to a conventional lever system. Chameleons and squid have independently evolved prey-capture structures that take advantage of this principle. By contracting transverse muscles in its tongue (chameleon) or tentacle (squid), the appendage rapidly extends to reach out and grab prey. Squid tentacles, for example, can extend an additional 70% from their resting length in 1/30 s or less. For a deeper look at muscular hydrostats, see Vogel (2013, pp. 419–421), and for a recent review of research on the topic, see Kier (2012).

5.4 THE CONSEQUENCES OF SIZE

Above, we saw how size affects load carrying: an ant can carry several times its own weight, as opposed to a human who struggles to lift more than a fraction of her or his weight. This difference has nothing to do with muscle properties—ant muscles are not intrinsically stronger than ours—but is mainly due to the way length, area, mass, and force scale with body size over a wide range. Moreover, the range from ant to human is not all that great, compared to living organisms in general, or multicellular animals in particular. Vogel (1988) cites a length range of 10^8 from bacteria to whales, whereas just within animals, the length range from the smallest insects and crustaceans to the largest whale is approximately 10^5. Within groups of generally similar organisms, the length range is not quite as dramatic but still spans orders of magnitude. Insects and mammals both have length ranges of approximately 10^3: insects from 10^{-4} to 10^{-1} m and mammals from 10^{-1} to 10^2 m. How do these ranges compare with size change due to growth in individuals? Humans are probably on the low end of the growth range because human newborns are relatively large, rarely increasing much more than 4 or 5 times in length and 30 or 40 times in mass. Growth increases much greater than this are common among other animals.

5.4.1 Surface to Volume Ratio

The surface to volume ratio (S/V) may be the most fundamental scaling relationship in biology. It constrains a huge range of physiological and mechanical processes, e.g., gas exchange, temperature regulating ability, skeletal proportions, load carrying, and terminal velocity. Small animals such as hummingbirds and honeybees have a lot of surface area for their volume, whereas large animals such as elephants and tyrannosaurs have much less surface area per unit volume. If animals maintained geometric similarity, then we could easily calculate relative S/V values for different sized animals just based on their lengths (L), because surface area would be proportional to L^2 and volume (and mass for living organisms) would be proportional to L^3. Although geometric similarity or "isometry" is sometimes assumed for first-order approximations, real animals never come very close to geometric similarity over any substantial size range. Consider a 5-mm long ant with a body mass of 3 mg versus a 30-m long blue whale. The blue whale is 6000 times longer, so by geometric similarity, $(6000)^3 \times 0.003$ g $= 6.5 \times 10^8$ g or 650 tonnes. In fact, an actual 30-m long blue whale has a body mass of approximately 180 tonnes, rather far off from geometric similarity.

Nevertheless, S/V matters even without exact geometric similarity. Ants can lift several times their body weight, not because they have unusually strong muscles, but because muscle force is related to cross-sectional area instead of mass, and as animals get smaller, their mass decrease much faster than their cross-sectional area: the ant simply has a lot more muscle cross section per unit mass than a person. Conversely, the low S/V of very large animals means that large dinosaurs such as *Tyranosaurus rex* were probably endothermic ("warm blooded") even without any insulation or elevated metabolic rate, simply due to their great size (Seebacher, 2003).

Skeletal support systems are greatly influenced by S/V. Galileo is usually credited with being the first to point out the difference in proportion of leg bones of large and small mammals. If an animal doubles in length, its weight increases by a factor of 8, and if its leg bones enlarged isometrically, they would have four times the cross-sectional area. So such geometrically similar leg bones, with four times the area but carrying eight times the load, would experience doubled stress. In fact, as Galileo pointed out (Schmidt-Nielsen, 1984, p. 43), when animals get larger, their leg bones get disproportionately larger in diameter to withstand greater loads: the leg bones of an animal that is 8 times heavier should have 8 rather than 4 times the cross-sectional area, or $(8)^{1/2} \approx 2.83$ times rather than 2 times greater diameter. When comparing a picture of a domestic cat's skeleton to that of a tiger, even with no indication of scale you can easily tell which is which, due to the more robust limb bones of the tiger. In fact, however, the cross-sectional area of mammalian skeletons do not quite compensate exactly for increasing mass (Vogel, 1988, p. 51), meaning that very large mammals seem to operate with relatively more fragile skeletons than their smaller counterparts.

At the other extreme, as animals get very tiny, their legs get increasingly spindly. Mass decreases much faster than body length, so the cross-sectional area needed for weight support becomes disproportionately low. The dimensions of the legs of insects and spiders demonstrate this phenomenon. The exoskeleton of an ant, scaled up to the size of an elephant, would collapse under its own weight (1950s-era grade-B science fiction movies notwithstanding). Few animals exemplify this better than opilionid[c] harvestmen or "daddy-longlegs." Some have legs that are 1 or 2 cm long, yet appear to have diameters not much wider than a human hair. Such proportions dictate a very low body mass and thus place strict limits on body size.

Among land animals, arthropods are very much on the small end of the size scale, whereas vertebrates are on the large end, and the two groups show surprisingly little overlap. Arthropods—insects, spiders, centipedes, etc.— have exoskeletons, in contrast to vertebrates that have endoskeletons, and both groups seem to devote about the same fraction of their body mass to their skeletons (Anderson et al., 1979). At least some researchers attribute the size differences to skeletal type. Currey, for example, compared the mechanical properties of endoskeletons and exoskeletons (Currey, 1967). He came to the conclusion that exoskeletons are inherently better at resisting large-scale, static loads, such as those that might cause bending or buckling—putting the skeleton on the outside gives it larger second moment of area. In contrast, endoskeletons are better at resisting impact loading, especially localized impacts, due to shape and thickness and also due to the cushioning by soft tissues surrounding the endoskeleton. At small sizes, exoskeletons can be strong enough to resist impacts, but at larger sizes, making an exoskeleton strong enough to resist impacts makes it prohibitively heavy. The largest arthropods—Maine lobsters, king crabs—are all marine, where underwater buoyancy helps offset the weight of a heavy, calcified exoskeleton; they are largely immobile in air. The fact that few vertebrates approach the small size of even large insects probably has less to do with endoskeletons being poorer at resisting large-scale loads and more to do with the fact that insects were already present, diverse, and highly successful by the time the first vertebrates ventured onto land: those "very small animal" ecological niches were already thoroughly filled.

---

c. Daddy-longlegs or harvestmen, order Opiliones, are not true spiders but are eight-legged arachnids related to spiders.

For an insightful (and entertaining) discussion of the consequences of size, scaling, surface-to-volume ratio, and allometry, see Chapter 3 of Vogel (1988), which is essentially repeated as Appendix 3 in Vogel (2013).

5.4.2 Maximum Jump Heights

Scientists have long known of a scaling argument—variously attributed to Galileo (Haldane, 1927) and Giovanni Borelli (Vogel, 2013)—that says that all animals should be able to jump to approximately the same absolute height. In other words, if we just look at how far they raise their center of mass, the absolute height should be about the same regardless of size (Schmidt-Nielsen, 1984). The general argument is that muscle work output is proportional to muscle mass, so if muscle makes up about the same fraction of the body mass in all animals (it does, usually approximately 40%–45%), an animal's muscle energy output will scale directly with mass. The same energy per unit mass will lift a body to the same height. Therefore all animals should have about the same jump height.[d] Do they?

As Table 5.1 shows, absolute jump height is surprisingly constant, given the enormous size range—over eight orders of magnitude—in body mass. This constancy is all the more amazing if jump height is expressed in relative height, i.e., jump height expressed in body lengths. In these terms, humans are rather pathetic, with a maximum height of 0.3 body lengths, whereas the

TABLE 5.1 Jump Heights of Animals Spanning a Range of Body Sizes

Animal	Mass (kg)	Jump Height (m)	Relative Jump Height	Takeoff Acceleration (g)
Antelope	200	1.2	1.0	1.6
Human	70	0.6	0.3	1.0
Domestic cat	2.5	1.5	5	3.2
Tree frog	0.013	0.65	35	6.1
Grasshopper	0.003	0.59	60	15
Flea	0.00049	0.2	130	245

Acceleration measured in units of gravities, g, where $1 g = 9.81$ m s^{-2}.
Data from Schmidt-Nielsen, K., 1984. Scaling: Why Is Animal Size so Important? Cambridge University Press, Cambridge, UK, 241 pp. and Biewener, A.A., 2003. Animal Locomotion. Oxford University Press, Oxford, UK, 281 pp.

d. A.V. Hill (1950) derived a similar result as part of an analysis of scaling of locomotion speed, using a more sophisticated geometrical argument.

smaller animals look increasingly impressive—tree frogs jump to a maximum height of about 35 body lengths, grasshoppers to about 60 body lengths, and fleas to an amazing 130 body lengths.

The range in absolute jump heights (Table 5.1) is fairly small, but in fact jump height is not completely independent of body size. Within a group of related species, larger individuals can usually jump to greater heights than smaller ones (Marsh, 1994). Also, longer limbs give jumping muscles longer time in contact with the ground and hence, greater takeoff speed, so longer-limbed animals tend to jump higher. Consider animals specialized for jumping, such as frogs, grasshoppers, and kangaroo rats—all have disproportionately long jumping legs.

Note that the smallest animals require a huge acceleration to reach the takeoff speed needed to achieve their measured jump heights. Muscles, in fact, cannot contract fast enough to produce those accelerations. The grasshopper reaches takeoff speed in 25 ms and the flea, in 0.8 ms (Schmidt-Nielsen, 1984). Intrinsic muscle speed (v_{IS}, lengths per second) does increase with decreasing size; recall that v_{IS} is inversely proportional to (body length)$^{\frac{1}{2}}$. To produce high enough accelerations at very small body sizes, however, v_{IS} would have to scale inversely with the first power of body length—the increased muscle speed as size decreases does not keep up with the needed acceleration. Even for such tiny animals as fleas, their muscles would need at least an order of magnitude more time to generate peak force, so how do these insects achieve such enormous accelerations?

Jumping insects almost uniformly use some sort of spring-and-catch arrangement to jump. They use muscles to compress a cuticular structure— usually containing resilin—held by a catch mechanism of some kind, and then use the elastic release of stored energy to produce the forces and accelerations needed to jump. Fleas, for example, have a pad of resilin at the base of each hind leg that is compressed over the course of 1/10 s or so against a catch. When the catch is released, the resilin spring accelerates the flea to jumping speed in less than 1 ms (Bennet-Clark and Lucey, 1967; Sutton and Burrows, 2011). Grasshoppers use a similar mechanism, although the spring in this case is the apodeme (tendon-like cuticular strut) of the main extensor muscle, as well as parts of the "knee" articulation (Bennet-Clark, 1975); resilin is present in both the catch mechanism as well as the spring itself (Burrows, 2016). The muscle contracts to store energy in these cuticular structures over at least 350 ms, but they extend the leg in only 25−30 ms when the catch is released (Bennet-Clark, 1975).

Another hurdle faced by small jumpers is air resistance, i.e., drag. Because of their small size, insect jumpers operate at such low Reynolds numbers that viscous drag becomes significant, at the same time their high S/V means that they have low inertia relative to their surface area. The air resistance faced by a jumping antelope or human is negligible, but that faced by a jumping insect can be substantial.

In the absence of air resistance, basic physics gives the vertical velocity needed at takeoff to achieve a particular jump height, so Bennet-Clark and Alder (1979) used an adjustable spring catapult to launch a variety of insects at a range of vertical speeds and measured their actual jump heights in air. They then calculated "jumping efficiency," h_a/h_v, where h_a is jump height in air and h_v is the height the jump would have reached in vacuum. They found that grasshoppers had jump efficiencies of approximately 0.8 at takeoff speeds that gave $h_v = 1.0$ m, whereas fleas with the same takeoff speed (and hence, same h_v) had jump efficiencies of only 0.4. In other words, for the same takeoff speed—same specific energy expenditure—a cat could jump to a height of 1.0 m, a grasshopper to 0.8 m, and a flea to 0.4 m. Because the drag will be highest at the highest speed, i.e., at the instant of takeoff, jumping efficiency actually increases as speed goes down: a flea with a takeoff speed sufficient to give it $h_v = 0.25$ m would have a jump efficiency of over 0.8, thus jumping to an actual height of over 0.2 m in air (and wasting less of its effort to overcome drag) (Bennet-Clark, 1980). From Table 5.1, we can see that in practice, air resistance is negligible for vertebrates, significant for large insects, and a serious impediment to jumping in small insects.

5.4.3 Growing Into Different Mechanical Realms?

Unlike other flying animals, pterosaurs appear to have been capable of powered flight throughout most of their juvenile development, over a wide range of body sizes (Bennett, 1996). Birds, bats, and insects either fly only as adults or begin flying only after they have reached nearly adult size. Alexander (2015, pp. 157−158) speculated that the size difference between very young and adult pterosaurs might entail differences in aerodynamics, due to operating at different Reynolds numbers. How different were they?

Pterosaurs are known only from fossils, so comparing body proportions among pterosaurs can be daunting. Nevertheless, wingspan is a useful characteristic for which reasonable data are available. Bennett (1996) considers all the *Rhamphorhynchus* specimens from the Solnhofen limestone to be a single species, and the humerus (upper arm bone) of the largest is 4.6 times longer than that of the smallest. The adults had wingspans of approximately 1.8 m, so as a first approximation, scaled to humerus length, young but flight-capable individuals might have had wingspans of 0.39 m. Based on data for birds with wingspans similar to these values (Tennekes, 2009), I calculate Reynolds number values from 30,000 to 42,000 for the smallest pterosaur versus 170,000 to 470,000 for the largest one. Although these values look quite different, even the lowest would still be considered "high Reynolds number" by biomechanics researchers, so young and old pterosaurs are not likely to operate under significantly different aerodynamic regimes: viscous drag should be negligible and turbulence should be similar for both. Given the typical change in body proportions with growth among vertebrates, the adult's wings were probably not exactly 4.6 times longer than the juvenile's, but even a change of 10% or 20% in lengths would not be enough to change the overall conclusion.

Operating at a very similar Reynolds number does not, however, mean that size did not affect flight. Assuming approximate geometric similarity, and based on scaling data from living birds, the small individuals would have weighed approximately 40 times less than the largest ones (Tennekes, 2009). Over such a size range, the small ones would probably have had to flap their wings about three times faster (Nudds et al., 2004). These differences would have had consequences for endurance, maneuverability, takeoff speed and angle, and soaring ability, among many other flight-related traits.

Plants, trees in particular, may illustrate a clearer example of growing from one mechanical regime into another. Even the simplest land plants, such as mosses, have a relatively complicated structure, consisting of a central mass of parenchyma cells surrounded by an epidermis of thicker-walled cells. Vascular plants add vessels and other mechanically supportive tissues, which perform a variety of physiological and mechanical functions.

The basic structural unit of a land plant is a cell that has a cell wall helically reinforced with cellulose fibers and is internally pressurized by osmotic processes. Biologists call this osmotic pressure *turgor pressure*, because the cells are turgid when pressurized. Many plants take advantage of the helical reinforcing fibers to direct growth. Herbaceous plants tend to do most of their growing at the tip of the shoot (a region called the "apical meristem") and cells that form from division at the apical meristem tend to have cell walls with fibers at a high fiber angle. Thus, when they are pressurized, they tend to elongate (and get narrower). In contrast, if plants are mechanically agitated, cells in the stem lay down cell wall layers with fibers at a low fiber angle. When these cells are pressurized, they tend to expand in diameter (and shorten) (Ennos, 2012, pp. 117—118).

The simplest arrangement of tissues found in land plants—thought to be similar to ancestral plants and today found mainly in reproductive stalks of mosses—is a stalk mostly made up of parenchyma cells surrounded by a thin outer layer of epidermal cells. The parenchyma cells are globular and tightly packed, whereas the epidermal cells are elongated and arranged parallel to the stalk's long axis. Both cell types are osmotically pressurized, which puts the parenchyma in compression and the epidermis in tension. How tall can such a hydrostatically stiffened column get?

Niklas (1992, p. 491) used an analysis developed for studying large deflections of beams and columns (the elastica theory) to calculate how tall a simple parenchyma—epidermis stem can get before it deflects more than 10 degrees under its own weight. Using actual values for the mechanical properties of parenchyma and epidermis, he calculated a maximum height of 55 cm for a stem of radius 0.9 cm (1.8 cm diameter), giving a length:diameter ratio (l/d) of approximately 30.

Almost all modern land plants—ferns, conifers, flowering plants—are vascular, that is, they have a variety of specialized tissues for transporting water and nutrients around the plant. Vascular plants have several other kinds of tissue in addition to parenchyma and epidermis, and the types that make up the transport structures can have dramatic effects on their mechanical properties. Herbaceous plants have primary vascular tissue as well as a fibrous supporting tissue called sclerenchyma. These tissues all form by differentiation from the apical meristem at the growing tip of the shoot, and they can both stiffen and strengthen the plant's structure. Niklas (1992, pp. 330–332) applied the elastica theory analysis to the flowering stalks of garlic; they are cylindrical with very little taper, which simplifies the analysis. Using values of E for actual garlic tissue, he calculated that with a typical radius of 0.3 cm, garlic stalks could achieve heights of up to 0.92 m before risking damage from self-loading. The tallest real garlic stalks he measured were only 0.51 m tall, which is quite reasonable given that the analysis only assumed static loading—real garlic stalks must account for wind, storms, pollinators landing, etc. So even with vascular tissue, herbaceous plants are fairly limited in height, although reaching l/d values of at least 170.

Several different lineages of plants—lycopods, horsetails, conifers, flowering plants—have evolved secondary growth, which can greatly increase plant size. Secondary growth occurs when a layer of cells just under the surface of a plant's stem divide and grow. Such growth allows plants to get taller in two ways. First, it causes an increase in diameter; if nothing else changed, greater diameter allows a stem to get taller before reaching its critical l/d. The other way that secondary growth increases plant height is, however, more important. One of the types of tissues formed by secondary growth is secondary xylem. Secondary xylem is the botanical term for wood. When secondary xylem first forms, it functions in water transport ("sapwood"), but as new layers are added to the periphery, the older layers give up their transport function and become the primary structural support (heartwood) of the plant. Secondary xylem fibers are nonliving, being composed mainly of cell walls of dead cells. The cell walls are very thick relative to the internal space where the living part of the cell was. The walls consist of several layers of helically wound cellulose fibers thickened and reinforced with the water-resistant protein lignin.

Whereas a hydrostatically supported cylindrical stem (pure parenchyma) can only achieve an l/d of about 15 before starting to fail under its own weight, a stem made of wood can support its own weight at an l/d of up to 600, 40 times higher (Niklas, 1992, p. 310). So between greatly increased diameter and much stronger material, secondary growth is largely responsible for the evolution of tall trees. For example, California redwoods include some of the tallest known trees. The tallest publically documented redwood so far discovered is an astonishing 115.6 m (380 ft) tall, but several dozen have been measured to be over 100 m (approximately 330 ft) tall. (Giant sequoias, close relatives of redwoods, do not get quite as tall but have wider trunks and so include the heaviest living organisms on the planet; the largest has an estimated mass of 1900 metric tonnes.) Although few other tree species approach such extreme heights, mature forest trees of all types still reach impressive heights. Tropical rainforest canopies range from 30 m (100 ft) to 45 m (150 ft) high, and individual trees of over 60 m (almost 200 ft) are common in Southeast Asian rainforests (Dudley and DeVries, 1990). The latter is taller than a 15-story building.

Secondary xylem thus allowed plants to move from the rather limited heights provided by hydrostatic tissues to the immense heights achieved by many species of trees. Moreover, most trees grow from seeds—weighing several orders of magnitude less than a mature tree—from which a tiny seedling grows. Tree seedlings thus start out in a size range that could easily be supported hydrostatically, but even young saplings quickly grow to heights (and l/d ratios) that would be impossible without secondary growth and woody tissue.

5.5 THE PROMISE OF BIOMIMICRY: HAVE WE ARRIVED?

As far back as 1990, Julian Vincent included an entire chapter on biomimicry in his book *Structural Biomaterials* (Vincent, 1990): the idea that biological materials or structures can provide a basis for the design of useful human-made products. The reasoning behind biomimicry goes that evolution often produces specialization and optimization, so if engineers need to carry out some process similar to something done by an organism, perhaps the organism has already evolved an efficient material or structure that can be copied. For example, bivalve mollusks called mussels secrete bundles of fibers called byssal threads that they attach to rocks with a tenacious, permanent adhesive that works under water. Geckos and house flies can walk up vertical glass walls and even walk upside down on glass ceilings. Bone tissue remodels itself automatically to increase strength where loads are highest. Researchers have long suggested that imitating or borrowing from such biological systems might lead to useful products. A handful of cases of successfully borrowing from nature—barbed wire based on thorny hedges, Velcro based on cockleburs, using wood (instead of cotton or linen) for paper based on paper wasp nests (Vogel, 1998)—may give the impression that if we just look hard enough, we can derive all sorts of useful designs from living organisms. Until recently, however, biomimicry as a modern research topic seemed to promise more than it delivered. At first such research was scattered across a wide variety of biological and engineering journals, but in 2006 the journal *Bioinspiration and Biomimetics* began publishing articles. This journal helped bring the concept of biomimicry to a much broader audience of researchers while promoting this new field.

5.5.1 Ornithopters

One of the earliest and the most touted areas of biomimicry was the development of *ornithopters*, aerial vehicles that fly by flapping their wings. Designs for large ornithopters go back at least to Leonardo da Vinci, but more recent researchers have focused on tiny vehicles of a similar size scale to flying animals. Around the turn of the 21st century, the Defense Advanced Research Projects Agency (DARPA) began funding researchers to develop microaerial vehicles (MAVs). DARPA's goal was to produce an autonomous aerial vehicle with a wingspan of less than 6 inches (15 cm), the ability to carry a camera, and 2 h of endurance (Alexander, 2009). Several research groups took on the challenge, and although none met all the goals—battery and digital camera technology of the day were not up to the task—they did produce a number of very small radio-controlled or autonomous flying machines. Some researchers went the route of craft with tiny propellers, or even tried to develop thumb-sized turbine (jet) engines, but others developed small ornithopters. Based on Froude efficiency (Chapter 3), bigger propellers are more efficient; they accelerate a lot of air a little, so using the whole wing as a propeller should, in principle, be much more efficient than a tiny, conventional propeller, especially at such small sizes. Although the MAVs that came out of the DARPA projects did not meet all the goals, they did inspire a new generation of engineers who became interested in the prospect of small, flapping-wing flying machines, which is to this day a very active area of research (e.g., Gerdes et al., 2012; Liu et al., 2016; Shyy et al., 2016; Chirarattananon et al., 2017). In a tour de force of reverse engineering and microfabrication, at least two different labs have produced and flown insect-scale MAVs based on house fly flight mechanics: the Micromechanical Flying Insect of Ron Fearing's lab at the University of California Berkeley (Wood et al., 2008) and the RoboBee of Robert Wood's lab at Harvard University (Chirarattananon et al., 2014). Although they can both fly, neither can carry an on-board power source, so both depend on tethers to the ground for power (and despite its name, the RoboBee is based on fly wings rather than bee wings).

5.5.2 Adhesives

Adhesive development is another area long seen as fertile ground for biomimicry. Many organisms attach themselves, temporarily or permanently, to solid objects. Some, such as mussels, barnacles, and oysters, do it underwater, in ways that might be very useful if we could emulate them. Others, such as house flies and geckos, seem to walk with equal ease on level, vertical, and even upside-down surfaces. Researchers have been particularly focused on gecko toe pads: a gecko's ability to walk upside down on both smooth and rough surfaces requires something that will stick to any kind of surface strongly enough to hold the gecko's entire weight, but that the gecko can easily detach to take a step. Gecko feet have no wet or sticky secretions. Instead, gecko toe pads are covered with very dense rows of very short, fine bristles called *setae*. Each bristle forms dozens or hundreds of very fine branches at its free end, and each branch ends in a nanoscopic[e] flat rectangle called a *spatula*. The best current evidence suggests that these millions of spatulae conform very closely to the substrate and stick due to van der Waals attraction. Imagine a product with the same properties as gecko toe pads: a tenacious but easily detachable, reusable adhesive (that also happens to be self-cleaning!). After much research on gecko feet, multiple research groups have developed prototype tapes with geckolike adhesive (Brodoceanu et al., 2016), although commercial scale manufacturing faces many hurdles. Readers should be aware that at least one brand of tape has been marketed with "gecko" in the name but which is nothing more than a type of duct tape.

5.5.3 Legged Robots

Locomotion has long been a popular area of biomimicry research. In its first year of publication, *Bioinspiration and Biomimetics* had articles on the efficacy of swimming with one pair versus two pair of flapping flippers by a swimming robot (Long et al., 2006); computer modeling of insect flight as a tool for developing MAVs (Żbikowski et al., 2006); computer modeling of undulatory swimming as a tool for developing undulatory robots (Eldredge, 2006); and a controller that allowed a small, autonomous glider to perform dynamic soaring like an albatross (this last example even used a process mimicking natural selection to refine the control algorithm) (Barate et al., 2006).

Legged locomotion in particular has been of special interest to roboticists because of its promise of easily traversing broken ground (McMahon, 1984, pp. 183–186). Researchers have developed many quadrupedal and hexapedal robots, and both mechanical and control systems have seen immense improvements. Until very recently, the promise of such systems for crossing broken terrain tended to outweigh the complexity of multilegged locomotion: exploratory vehicles sent to Mars and the moon have all used wheels. That could change, as researchers have now demonstrated a quadrupedal robot that can "gallop" (it actually bounds) and leap over barriers of the robot's hip height while galloping (MIT Cheetah) (Chu, 2015), and another that can walk on steep, muddy paths or on uneven snow and ice and can successfully catch itself and keep from falling when it slips on ice (Boston Dynamics BigDog) or is pushed vigorously by a researcher (Raibert et al., 2008).

Bipedal locomotion adds another layer of control complexity: a bipedal robot needs a fair amount of sensory feedback to maintain balance and avoid

---

e. "Nanoscopic" in technology refers to objects on the size scale of 1–100 nm, far smaller than can be seen under a light microscope, visible only with more powerful instruments such as electron microscopes.

falling over. Nevertheless, quite a bit of research has gone into bipedal mechanical locomotion for two main uses: first, as part of designing more humanoid robots and second, to produce more effective prosthetic limbs for lower-limb amputees (Torricelli et al., 2016). Researchers have even built bipedal robots that can run on uneven surfaces, although most require external power and tethers to prevent damaging falls. As miniaturization proceeds and battery technology advances, practical untethered bipedal robots are currently being developed.

After a couple of decades of more promise than production, biomimicry has finally begun to pay off. Fishlike robots, tiny ornithopters, and legged robots have become mature technology and are now just starting to move out of the lab and into commercial use. Researchers have developed several gecko-based adhesive materials, and tapes using such material should soon be commercially available. Panels and coatings using "riblet" technology for drag reduction—based on the fine structure of hummingbird feathers and shark skin—are in commercial production and have been applied to boat hulls and airliner fuselages.

People sometimes have a tendency to think that "natural" is automatically better than "artificial" and by extension, assume that products based on biology should be inherently superior. This tendency is biased, and often, not supported by facts. What biomechanics researchers and engineers must keep in mind is that, yes, evolution does tend to refine and even optimize systems, but natural selection has a severely limited selection of materials from which to choose. Organisms cannot make body parts out of steel, or Kevlar, or granite. If a decade or more of active, widespread, bio-inspired design research has shown anything, it is that just because a biomimetic design is elegant or efficient does not necessarily make it more cost-effective or more practical than a conventional design. For functions similar or analogous to those encountered by animals and plants, however, sometimes a bio-inspired design might win out over a conventional design.

FURTHER READING

Biomimicry

Brodoceanu, D., Bauer, C.T., Kroner, E., Arzt, E., Kraus, T., 2016. Hierarchical bioinspired adhesive surfaces—a review. Bioinspiration and Biomimetics 11, 051001. http://dx.doi.org/10.1088/1748-3190/11/5/051001.

Gerdes, J.W., Wilkerson, S.A., Gupta, S.K., 2012. A review of bird-inspired flapping wing miniature air vehicle designs. Journal of Mechanisms and Robotics 4, 021003. http://dx.doi.org/10.1115/1.4005525.

Kapsali, V., 2016. Biomimicry for Designers: Applying Nature's Processes and Materials in the Real World. Thames & Hudson, London, 240 pp.

Vincent, J.F.V., 1990. Structural Biomaterials. Princeton University Press, Princeton, New Jersey, 206 pp.

Hydrostatic Skeletons

Ennos, A.R., 2012. Solid Biomechanics. Princeton University Press, Princeton, New Jersey, 264 pp.

Kier, W.M., 2012. The diversity of hydrostatic skeletons. Journal of Experimental Biology 215, 1247–1257.

Maximum Jump Heights

Bennet-Clark, H.C., Alder, G.M., 1979. The effect of air resistance on the jumping performance of insects. Journal of Experimental Biology 82, 105–121.

Sutton, G.P., Burrows, M., 2011. Biomechanics of jumping in the flea. Journal of Experimental Biology 214, 836–847.

Muscles and Locomotion

Goldspink, G., 1980. Locomotion and the sliding filament mechanism. In: Elder, H.Y., Trueman, E.R. (Eds.), Aspects of Animal Locomotion. Cambridge University Press, Cambridge, UK, pp. 1–25.

van Leeuwen, J.L., 1992. Muscle function in locomotion. In: Alexander, R.M. (Ed.), Mechanics of Animal Locomotion. Springer-Verlag, New York, pp. 191–250.

Vogel, S., 2001. Prime Mover: A Natural History of Muscle. W.W. Norton & Company, New York, 370 pp.

Size and Scaling

Schmidt-Nielsen, K., 1984. Scaling: Why Is Animal Size so Important? Cambridge University Press, Cambridge, UK, 241 pp.

Vogel, S., 1988. Life's Devices: The Physical World of Animals and Plants. Princeton University Press, Princeton, New Jersey, 367 pp.

Terrestrial (Legged) Locomotion

Alexander, R.M., 2003. Principles of Animal Locomotion. Princeton University Press, Princeton, New Jersey.

Biewener, A.A., 2003. Animal Locomotion. Oxford University Press, Oxford, UK, 281 pp.

McGreer, T., 1992. Principles of walking and running. In: Alexander, R.M. (Ed.), Mechanics of Animal Locomotion. Springer-Verlag, New York, pp. 114–139.

Chapter 6

Organismal Versus Technological Design

6.1 BORROWING FROM ENGINEERS

Throughout this book, we have used concepts borrowed from human technology, mainly mechanical engineering, to describe and explain the mechanical properties of organisms. The fundamental concepts developed for engineering mechanics are largely fundamental to biomechanics as well. If biomechanists had been forced to start from scratch—without the underlying principles developed over two centuries by applied physicists and engineers—our knowledge of the mechanics of the biological world would surely be a drop in the bucket compared with our actual knowledge, that is, if the field even existed at all. Yet, as we have seen, nature tends to apply those principles in very different ways from human designers.

The whole concept of biomimetics or biomimicry, as described in the previous chapter, is based on the idea that biomechanics can provide new mechanical concepts to form the basis for novel technological designs. A handful of examples of such "technology transfer" from nature do exist. As has been noted elsewhere (Vogel, 1998, p. 18), however, the converse is more typical: the biomechanics researcher is more likely to notice some nifty mechanical property of an organism after the engineer has already discovered and used the property. For example, the leading-edge vortex was first described on the wings of insects in 1996 (Ellington et al., 1996) but had been known to engineers studying delta-winged aircraft for decades (Bertin and Smith, 1979, p. 208).

This chapter's title suggests that an organism can have a design, which might be taken to imply the existence of a designer. When biologists speak of an organism's "design," they are referring to the combination of features that allow that organism to perform various tasks and functions and that may be well-suited to particular habitats or situations. A tuna's low drag body shape, fatigue-resistant swimming muscles, and hydrodynamically efficient tail make it a fast, efficient swimmer, so it has a design appropriate for high-speed, long-distance swimming. Those features are not "designed" in the sense used in technology; they are produced by the refining effects of natural selection. Variation in nature is largely random, being the product of random mutations,

Nature's Machines. http://dx.doi.org/10.1016/B978-0-12-804404-9.00006-2

151

and almost all such changes are either neutral[a] or harmful, and the latter are weeded out by natural selection. Only a very tiny fraction of such mutations produces a beneficial change that improves survival or reproduction. The design process in nature is thus the result of the vast majority of novel features being discarded, whereas the extremely rare beneficial novelty spreads through the population because of the advantage it confers. We can thus refer to an organism's "design" and compare it to technological designs, without implying the necessity of a designer for the organism (Vogel, 1998, p. 22; Alexander, 2009, pp. 13−14).

6.2 DIFFERENT MATERIALS USED IN DIFFERENT WAYS

In spite of the fact that biomechanics researchers have borrowed fundamental concepts lavishly from engineers, nature has clearly applied those principles in very different ways. Earlier authors have made the comparison between the mechanics of nature and the mechanics of engineers, sometimes in considerable detail (e.g., Vogel, 1988, 1998). A few of the more dramatic divergences are given below.

6.2.1 Materials

Human technology tends to use materials that are dry and stiff. Some are strong, such as metals, whereas others are brittle, such as glass and masonry. Organisms tend to use materials that are wet and pliant. Some are weak or brittle, such as mesoglea, whereas others are tough, such as collagen or insect cuticle. Both domains make use of ceramics, but human-made ceramics are usually dry and inorganic and thus often brittle, such as pottery. Biological ceramics are moist and infiltrated with protein and thus tougher, such as bone and mollusk shells.

The biggest difference may be the complete lack of metals—in metallic form—seen in organismal structures. Organisms use many metal elements, but always as ions or individual atoms in complex molecules, such as the four iron atoms in each of our hemoglobin molecules. Biologists have yet to discover an organism that makes structural use of any metal. Pure metals and metallic alloys, in contrast, underpin most of our modern technology. Civilization would collapse—literally and figuratively—without steel and iron.

6.2.2 Shape

Engineers and builders make abundant use of straight edges, flat surfaces, and right angles. Human-made structures are very often in the form of cubes or

---

a. Due to redundancies in the genetic code, a point mutation, i.e., a change in a single DNA base pair, may not change the meaning of the 3-base codon or genetic "word"; if not, such a change would be a neutral mutation.

boxes. When cylinders or spheres are used, they are usually for specific purposes, such as pipes, shafts, or bearings, and rarely as major structural components. Flat plates, sharp corners, and straight lines are rare in nature; indeed, human designs that incorporate curved edges and surfaces are often described as "organic." Cylinders and spheres are extremely common as both major structural elements and smaller-scale components of organisms. The difference may partly reflect fabrication methods—carpentry and bricklaying versus growth—and partly the fact that many organic structures are pressurized, for which boxes with flat sides are ill suited.

6.2.3 Loading and Movement

Although may exceptions exist, humans tend to build large structures that are loaded more in compression than tension, whereas the bodies of organisms tend to be arranged more to resist tension. Perhaps the clearest illustration of the tensile character of life is nature's plentiful use of pressurized hydrostats. Not only are such hydrostats common among both animals and plants, they almost always use water as the pressurizing fluid. Humans use few pressurized structural components and those few are almost all pressurized with air or some other gas (and are thus technically "aerostats"). Examples include blimps[b]—pressurized with helium—and automobile tires.

The difference in materials leads to difference in connections and joints. Because we make things out of hard, stiff material, we usually make joints with parts that slide or roll past each other. Although many vertebrate skeletal joints do use this mechanism, outside of vertebrate skeletons, movements between parts are usually based on bending or twisting. Venus fly traps close when the curved hinge line suddenly straightens; leaves on trees collapse together in wind due to twisty petioles. Sometimes organisms develop weak area intentionally, allowing a kink to form in a predetermined place to act as a hinge, as in some very tiny arthropod legs. Such techniques do not scale up well and so are not suited to most machinery, although the use of grooves in small plastic objects to form integral hinges—so-called "live" hinges—has become common for very light duty uses in our technology.

Engineers use struts and joints very differently from organisms. Combinations of struts and joints form lever systems, and in our technology, we usually use lever systems to amplify force at the expense of distance or speed. This amplification is the basis for pulling nails out of wood with a claw hammer or raising an automobile with a manual jack. In contrast, the lever systems formed by vertebrate or arthropod skeletons do exactly the opposite:

---

b. Not to be confused with dirigibles from the early 20th century: dirigibles such as Germany's *Hindenburg* and the US Navy's *Akron* had a rigid, internal supporting framework, whereas blimps, such as the lighter-than-air ships in current use, get their shape mostly or completely from being pressurized.

they greatly amplify the speed or distance of appendage movement at the cost of proportionately decreased force. A person's elbow might convert the biceps muscle's pull of 1 cm at a force of 75 N into a 15-cm movement of the hand at a force of 5 N. This situation is the exact opposite of "favorable leverage" in the technological sense.

The difference in terrestrial locomotion between technology and biology embodies many of the differences described above. Vertebrates and arthropods walk or run on jointed limbs, whereas human-built vehicles mostly roll on wheels. Animals use muscles, which do not allow continuous rotation—no animal has ever evolved a true rotating wheel or axle system as part of its body—so wheels are not an option for them, whereas the lever system of a jointed skeleton makes an effective locomotion mechanism. As others have pointed out, wheels are actually not of much use without roads (Labarbera, 1983). Roads are not completely essential for wheels, but the more uneven the terrain, the bigger the wheels need to be. The covered wagons—"prairie schooners"—of the 19th-century North American migrants used huge wheels, over 1.2 m in diameter, to travel off roads, yet even then the wagons were stopped by obstructions that a person or a horse could easily step over. Because we can manufacture wheels and axles, and because wheels make transporting loads so much easier, we invest great amounts of effort and money into building roads—yet another example of straight lines and flat surfaces in our technology.

6.2.4 The Construction Process

A new organism is "constructed" by the process of growth and development. These processes are guided partly by instructions encoded in genes and partly by feedback from the environment. Changes and innovations in organismal design are glacially slow because the vast, overwhelming majority of mutations, the main source of innovation, are either detrimental or have no effect, so are not spread through the population. Mutations that are beneficial are extremely rare and appear randomly and spontaneously. If they provide a significant advantage, they tend to spread through the population because those individuals with the mutation tend to survive and reproduce slightly better. Even such a beneficial mutation, however, takes many generations to become common throughout a population. Biological change is thus exceedingly slow and not goal directed.

Our technology operates in stark contrast to biology. An engineer or a designer generates a design with one or more specific tasks to perform. The process is inherently goal directed. Moreover, depending on the social environment, innovation can be very rapid. The development of aviation in the 20th century—from the Wright brothers to Mach 6 research airplanes and jumbo jets—illustrates how changes that might take millions of generations in nature can be compressed into the span of a handful of generations by the goal-directed, innovation-seeking nature of modern human technology.

6.3 RESEARCH AND METHODS[c]

Many studies of biomechanics in the mid to late 20th century used methods and equipment borrowed directly from engineers (e.g., Vosburgh, 1982). As research questions and theoretical analyses became more sophisticated, researchers realized that the wet, floppy, irregular bits of animals and plants that they studied were not well matched to traditional engineering equipment. Biomechanics researchers began to develop their own techniques. Some proved to be elegant in their simplicity, such as the weight hanging from a spiral pulley that Vogel and Papanicolaou (1983) used to apply a constant stress as the cross-sectional area changes in creep tests on viscoelastic material. Similarly, Farran et al. (2008) modified a standard universal testing machine with parts from a simple pair of toenail clippers to measure fracture properties of keratinous structures. Only slightly more complex, the portable toughness tester of Darvell et al. (1996) is designed specifically to test floppy or compliant materials in the type of field setting commonly faced by biologists. Sonomicrometry, using tiny piezo crystals to measure very small displacements acoustically, lets researchers measure soft tissue displacements in living animals (Biewener et al., 1998a). Flight in gas mixtures other than air—primarily helium and oxygen—has allowed researchers to test the limits of energetics and power output in hovering by hummingbirds (Altshuler and Dudley, 2003) and bats (Dudley and Winter, 2002).

Other techniques require more expensive equipment and computerized data analysis. Large strains of compliant structures such as skin can be measured by computer-aided optical correlation (Da Fonseca et al., 2005). In contrast, at the very low end of the strain spectrum, laser speckle interferometry allows researchers to measure strains of stiff materials such as teeth noninvasively and at smaller strains than permitted by traditional strain gauges (Zaslansky et al., 2006).

In the fluid realm, another laser-based technique, digital particle image velocimetry, uses laser sheets to illuminate neutral density particles in air or water that are assumed to follow the flow patterns. The particle density is too high to follow particles individually, so sophisticated autocorrelation or signal processing algorithms are used to extract flow patterns from successive, high-resolution digital video images. This allows reconstruction of very fine details of the flow pattern, all without any physical probes or flow disturbances (Santhanakrishnan et al., 2012; Crandell and Tobalske, 2015).

High-speed photography (nowadays normally video) has long been a mainstay of many areas of biomechanics (e.g., Alexander, 1986), and prices of high-speed video systems have decreased as ease of use and versatility have increased in recent years. For instance, based on kinematics measured from such movies or videos, researchers studying insect flight have developed a

---

c. Adapted from Alexander (2016).

series of scaled, dynamically similar physical models of flapping insect wings. These range from the "Robo-moth" of Ellington's lab that aided in the discovery of the leading-edge vortex (Ellington et al., 1996) to recent computer-controlled models of crane fly wings (Ishihara et al., 2009), hawk moth wings (Cheng et al., 2011), and house fly wings (Elzinga et al., 2014).

As a result of the ever-increasing computational power of computers, modeling of animal flight continues to become more detailed and realistic to the point that such models can predict forces and moments on the wings and bodies of hovering insects (Liu and Sun, 2008) with reasonable accuracy. Engineers have developed computational fluid dynamics (CFD) techniques based on numerical solutions of the Navier–Stokes equations to aid airplane design, and researchers have modified and extended CFD models to take into account the fundamentally unsteady nature of animal flight (Sun and Tang, 2002; Sun and Xiong, 2005; Yamamoto and Isogai, 2005). These CFD methods often require extreme computational resources (i.e., supercomputers). Sophisticated aerodynamic models based on observed features of the flow patterns can be computationally simpler than CFD models, yet they provide accurate predictions of flapping wing aerodynamics (Minotti, 2002). These advanced methods are often explicitly described as tools for designing microaerial vehicles that mimic insect flight (Żbikowski, 2002; Ansari et al., 2006). At the other end of the size spectrum, researchers have developed computerized tools derived from helicopter aerodynamics to analyze likely flight mechanics of giant extinct birds (Chatterjee et al., 2007) and the largest pterosaurs (Chatterjee and Templin, 2004).

Clever and observant biomechanists will probably continue to describe seemingly mundane situations that have startlingly unexpected mechanics. A case in point: Reis et al. (2010) showed that cats lap water using a technique that is completely unlike the conventional wisdom that they use their tongues as spoons. Along with examples throughout this book—kelp tougher than steel, helically reinforced shark skin—these surprising mechanical features remind us that the biomechanics of organisms is not always as obvious or straightforward as it may first appear.

FURTHER READING

Alexander, D.E., 2009. Why Don't Jumbo Jets Flap Their Wings? Flying Animals, Flying Machines, and How They Are Different. Rutgers University Press, New Brunswick, New Jersey, 278 pp.

Labarbera, M., 1983. Why the wheels won't go. American Naturalist 121, 395–408.

Reis, P.M., Jung, S.H., Aristoff, J.M., Stocker, R., 2010. How cats lap: water uptake by *Felis catus*. Science 330, 1231–1234.

Vogel, S., 1998. Cats' Paws and Catapults: Mechanical Worlds of Nature and People. W.W. Norton, New York, 382 pp.

Bibliography

Albert, R., Pfeifer, M.A., Barabasi, A.L., Schiffer, P., 1999. Slow drag in a granular medium. Physical Review Letters 82, 205–208.

Alexander, D.E., 1986. Wind tunnel studies of turns by flying dragonflies. Journal of Experimental Biology 122, 81–98.

Alexander, D.E., 1990. Drag coefficients of swimming animals: effects of using different reference areas. Biological Bulletin 179, 186–190.

Alexander, D.E., 2002. Nature's Flyers: Birds, Insects, and the Biomechanics of Flight. Johns Hopkins University Press, Baltimore, Maryland, 358 pp.

Alexander, D.E., 2009. Why Don't Jumbo Jets Flap Their Wings? Flying Animals, Flying Machines, and How They Are Different. Rutgers University Press, New Brunswick, New Jersey, 278 pp.

Alexander, D.E., 2015. On the Wing: Insects, Pterosaurs, Birds, Bats and the Evolution of Animal Flight. Oxford University Press, New York, 210 pp.

Alexander, D.E., 2016. The biomechanics of solids and fluids: the physics of life. European Journal of Physics 37, 053001. http://dx.doi.org/10.1088/0143-0807/37/5/053001.

Alexander, R.M., 1959a. The physical properties of the swimbladder in intact Cypriniformes. Journal of Experimental Biology 36, 315–332.

Alexander, R.M., 1959b. The physical properties of the isolated swimbladder in Cyprinidae. Journal of Experimental Biology 36, 341–346.

Alexander, R.M., 1962. Visco-elastic properties of body-wall of sea anemones. Journal of Experimental Biology 39, 373–386.

Alexander, R.M., 1968. Animal Mechanics. University of Washington Press, Seattle, 346 pp.

Alexander, R.M., 1982. Locomotion of Animals. Blackie & Son, London, 163 pp.

Alexander, R.M., 1983. Animal Mechanics, second ed. Blackwell Scientific Publications, Oxford. 301 pp.

Alexander, R.M., 2003. Principles of Animal Locomotion. Princeton University Press, Princeton, New Jersey, 371 pp.

Alexander, R.M., Bennett, M.B., 1987. Some principles of ligament function, with examples from the tarsal joints of the sheep (Ovis aries). Journal of Zoology 211, 487–504.

Alexander, R.M., Jayes, A.S., 1983. A dynamic similarity hypothesis for the gaits of quadrupedal mammals. Journal of Zoology 201, 135–152.

Alexander, R.M., Maloiy, G.M.O., Njau, R., Jayes, A.S., 1979. Mechanics of running of the ostrich (Struthio camelus). Journal of Zoology 187, 169–178.

Alexander, R.M., Vernon, A., 1975. The mechanics of hopping by kangaroos (Macropodidae). Journal of Zoology 177, 265–303.

Altshuler, D.L., Dudley, R., 2003. Kinematics of hovering hummingbird flight along simulated and natural elevational gradients. Journal of Experimental Biology 206, 3139–3147.

Altshuler, D.L., Princevac, M., Pan, H.S., Lozano, J., 2009. Wake patterns of the wings and tail of hovering hummingbirds. Experiments in Fluids 46, 835–846.

Anderson, J.F., Rahn, H., Prange, H.D., 1979. Scaling of supportive tissue mass. The Quarterly Review of Biology 54, 139–148.

Ansari, S.A., Żbikowski, R., Knowles, K., 2006. Aerodynamic modelling of insect-like flapping flight for micro air vehicles. Progress in Aerospace Sciences 42, 129–172.

Ar, A., Rahn, H., Paganelli, C.V., 1979. The avian egg: mass and strength. Condor 81, 331–337.

Bailey, S.W., 1954. Hardness of arthropod mouthparts. Nature 173, 503.

Bainbridge, R., 1958. The speed of swimming of fish as related to size and to the frequency and amplitude of the tail beat. Journal of Experimental Biology 35, 109–133.

Balashov, V., Preston, R.D., Ripley, G.W., Spark, L.C., 1957. Structure and mechanical properties of vegetable fibres. I. The influence of strain on the orientation of cellulose microfibrils in sisal leaf fibre. Proceedings of the Royal Society of London B: Biological Sciences 146, 460–468.

Barate, R., Doncieux, S., Meyer, J.-A., 2006. Design of a bio-inspired controller for dynamic soaring in a simulated unmanned aerial vehicle. Bioinspiration and Biomimetics 1, 76–88.

Barnett, C.H., 1958. Measurement and interpretation of synovial fluid viscosities. Annals of the Rheumatic Diseases 17, 229–233.

Batchelor, G.K., 1967. An Introduction to Fluid Dynamics. Cambridge University Press, Cambridge, UK, 615 pp.

Beer, F.P., Johnston, E.R., 1981. Mechanics of Materials. McGraw-Hill Book Co., New York, 616 pp.

Bennet-Clark, H.C., 1975. The energetics of the jump of the locust *Schistocerca gregaria*. Journal of Experimental Biology 63, 53–83.

Bennet-Clark, H.C., 1980. Aerodynamics of insect jumping. In: Elder, H.Y., Trueman, E.R. (Eds.), Aspects of Animal Locomotion. Cambridge University Press, Cambridge, UK, pp. 151–167.

Bennet-Clark, H.C., Alder, G.M., 1979. The effect of air resistance on the jumping performance of insects. Journal of Experimental Biology 82, 105–121.

Bennet-Clark, H.C., Lucey, E.C.A., 1967. The jump of the flea: a study of the energetics and a model of the mechanism. Journal of Experimental Biology 47, 59–76.

Bennett, S.C., 1996. Year-classes of pterosaurs from the Solnhofen limestone of Germany: taxonomic and systematic implications. Journal of Vertebrate Paleontology 16, 432–444.

Bennett, S.C., 2000. Pterosaur flight: the role of actinofibrils in wing function. Historical Biology 14, 255–284.

Bergel, D.H., 1961. Dynamic elastic properties of arterial wall. Journal of Physiology 156, 458–469.

Bertin, J.J., Smith, M.L., 1979. Aerodynamics for Engineers. Prentice-Hall, Englewood Cliffs, New Jersey, 410 pp.

Biewener, A.A., 2003. Animal Locomotion. Oxford University Press, Oxford, UK, 281 pp.

Biewener, A.A., Alexander, R.M., Heglund, N.C., 1981. Elastic energy storage in the hopping of kangaroo rats (*Dipodomys spectabilis*). Journal of Zoology 195, 369–383.

Biewener, A.A., Corning, W.R., Tobalske, B.W., 1998a. In vivo pectoralis muscle force-length behavior during level flight in pigeons (*Columba livia*). Journal of Experimental Biology 201, 3293–3307.

Biewener, A.A., Konieczynski, D.D., Baudinette, R.V., 1998b. In vivo muscle force-length behavior during steady-speed hopping in tammar wallabies. Journal of Experimental Biology 201, 1681–1694.

Blackledge, T.A., Hayashi, C.Y., 2006. Silken toolkits: biomechanics of silk fibers spun by the orb web spider *Argiope argentata* (Fabricius 1775). Journal of Experimental Biology 209, 2452–2461.

Blokhin, S.A., 1984. Investigations of gray whales taken in the Chukchi coastal waters. In: Jones, M.L., Swartz, S.L., Leatherwood, S. (Eds.), The Gray Whale: *Eschrichtius robustus*. Academic Press, Orlando, Florida, pp. 487–509.

Boettiger, E., Furshpan, E., 1952. The mechanics of flight movements in Diptera. Biological Bulletin 102, 200–211.

Bomphrey, R.J., Taylor, G.K., Thomas, A.L.R., 2009. Smoke visualization of free-flying bumblebees indicates independent leading-edge vortices on each wing pair. Experiments in Fluids 46, 811–821.

Brodoceanu, D., Bauer, C.T., Kroner, E., Arzt, E., Kraus, T., 2016. Hierarchical bioinspired adhesive surfaces—a review. Bioinspiration and Biomimetics 11, 051001. http://dx.doi.org/10.1088/1748-3190/11/5/051001.

Brown, R.H.J., 1948. The flight of birds: the flapping cycle of the pigeon. Journal of Experimental Biology 25, 322–333.

Brown, R.H.J., 1953. The flight of birds. 2. Wing function in relation to flight speed. Journal of Experimental Biology 30, 90–103.

Burrows, M., 2016. Development and deposition of resilin in energy stores for locust jumping. Journal of Experimental Biology 219, 2449–2457.

Caro, C.G., Pedley, T.J., Schroter, R.C., Seed, W.A., 2012. The Mechanics of the Circulation. Oxford University Press, Oxford, UK, 527 pp.

Cavagna, G.A., Heglund, N.C., Taylor, C.R., 1977. Mechanical work in terrestrial locomotion: two basic mechanisms for minimizing energy expenditure. American Journal of Physiology – Regulatory, Integrative and Comparative Physiology 233, R243–R261.

Chatterjee, S., Templin, R.J., 2004. Posture, locomotion and paleoecology of pterosaurs. Geological Society of America Special Papers 376, 1–64.

Chatterjee, S., Templin, R.J., Campbell, K.E., 2007. The aerodynamics of *Argentavis*, the world's largest flying bird from the Miocene of Argentina. Proceedings of the National Academy of Sciences of the United States of America 104, 12398–12403.

Che, J., Dorgan, K.M., 2010. Mechanics and kinematics of backward burrowing by the polychaete *Cirriformia moorei*. Journal of Experimental Biology 213, 4272–4277.

Chen, J., Friesen, W.O., Iwasaki, T., 2011. Mechanisms underlying rhythmic locomotion: body-fluid interaction in undulatory swimming. Journal of Experimental Biology 214, 561–574.

Cheng, B., Deng, X.Y., Hedrick, T.L., 2011. The mechanics and control of pitching manoeuvres in a freely flying hawkmoth (*Manduca sexta*). Journal of Experimental Biology 214, 4092–4106.

Chirarattananon, P., Chen, Y.F., Helbling, E.F., Ma, K.Y., Cheng, R., Wood, R.J., 2017. Dynamics and flight control of a flapping-wing robotic insect in the presence of wind gusts. Interface Focus 7, 20160080. http://dx.doi.org/10.1098/rsfs.2016.0080.

Chirarattananon, P., Ma, K.Y., Wood, R.J., 2014. Adaptive control of a millimeter-scale flapping-wing robot. Bioinspiration and Biomimetics 9, 025004. http://dx.doi.org/10.1088/1748-3182/9/2/025004.

Chu, J., 2015. MIT Cheetah Robot Lands the Running Jump. MIT News. Published by Massachusetts Institute of Technology. http://news.mit.edu/2015/cheetah-robot-lands-running-jump-0529.

Clark, B.D., Bemis, W., 1979. Kinematics of swimming of penguins at the Detroit Michigan USA Zoo. Journal of Zoology 188, 411–428.

Clark, R.B., Cowey, J.B., 1958. Factors controlling the change of shape of certain nemertean and turbellarian worms. Journal of Experimental Biology 35, 731–748.

Close, R.I., 1972. Dynamic properties of mammalian skeletal muscles. Physiological Reviews 52, 129–197.

Cox, R.G., 1970. The motion of long slender bodies in a viscous fluid. Part 1. General theory. Journal of Fluid Mechanics 44, 791–810.

Crandell, K.E., Tobalske, B.W., 2015. Kinematics and aerodynamics of avian upstrokes during slow flight. Journal of Experimental Biology 218, 2518–2527.

Cribb, B.W., Stewart, A., Huang, H., Truss, R., Noller, B., Rasch, R., Zalucki, M.P., 2008. Insect mandibles: comparative mechanical properties and links with metal incorporation. Naturwissenschaften 95, 17–23.

Cross, R.C., 2011. Physics of Baseball & Softball. Springer-Verlag, New York, 324 pp.

Currey, J.D., 1967. The failure of exoskeletons and endoskeletons. Journal of Morphology 123, 1–16.

Currey, J.D., 1977. Mechanical properties of mother of pearl in tension. Proceedings of the Royal Society of London B: Biological Sciences 196, 443–463.

Currey, J.D., 1980a. Skeletal factors in locomotion. In: Elder, H.Y., Trueman, E.R. (Eds.), Aspects of Animal Locomotion. Cambridge University Press, Cambridge, UK, pp. 27–48.

Currey, J.D., 1980b. Mechanical properties of mollusc shell. In: Vincent, J.F.V., Currey, J.D. (Eds.), The Mechancial Properties of Biological Materials, vol. 34. Society for Experimental Biology, Cambridge, UK, pp. 75–97.

Da Fonseca, J.Q., Mummery, P.M., Withers, P.J., 2005. Full-field strain mapping by optical correlation of micrographs acquired during deformation. Journal of Microscopy 218, 9–21.

Daniel, T.L., 1981. Fish mucus: In situ measurements of polymer drag reduction. Biological Bulletin 160, 376–382.

Daniel, T.L., 1983. Mechanics and energetics of medusan jet propulsion. Canadian Journal of Zoology 61, 1406–1420.

Daniel, T.L., 1984. Unsteady aspects of aquatic locomotion. American Zoologist 24, 121–134.

Daniel, T.L., Jordan, C., Grunbaum, D., 1992. Hydromechanics of swimming. In: Alexander, R.M. (Ed.), Mechanics of Animal Locomotion. Springer-Verlag, New York, pp. 17–49.

Daniel, T.L., Meyhofer, E., 1989. Size limits in escape locomotion of carridean shrimp. Journal of Experimental Biology 143, 245–265.

Darvell, B.W., Lee, P.K.D., Yuen, T.D.B., Lucas, P.W., 1996. A portable fracture toughness tester for biological materials. Measurement Science and Technology 7, 954–962.

Denny, M.W., 1980. Silks: their properties and functions. Symposium of the Society for Experimental Biology 34, 247–272.

Denny, M.W., 1981. A quantitative model for the adhesive locomotion of the terrestrial slug, *Ariolimax columbianus*. Journal of Experimental Biology 91, 195–217.

Denny, M.W., 1984. Mechanical properties of pedal mucus and their consequences for gastropod structure and performance. American Zoologist 24, 23–36.

Denny, M.W., 1988. Biology and the Mechanics of the Wave-Swept Environment. Princeton University Press, Princeton, New Jersey, 329 pp.

Denny, M.W., 1993. Air and Water: The Biology and Physics of Life's Media. Princeton University Press, Princeton, New Jersey, 341 pp.

Denny, M.W., Daniel, T.L., Koehl, M.A.R., 1985. Mechanical limits to size in wave-swept organisms. Ecological Monographs 55, 69–102.

Denny, M.W., Gosline, J.M., 1980. The physical properties of the pedal mucus of the terrestrial slug, *Ariolimax columbianus*. Journal of Experimental Biology 88, 375–394.

Dimery, N.J., Alexander, R.M., Ker, R.F., 1986. Elastic extension of leg tendons in the locomotion of horses (*Equus caballus*). Journal of Zoology 210, 415–425.

Downey, J.M., 2003. Hemodynamics. In: Johnson, L.R. (Ed.), Essential Medical Physiology. Academic Press/Elsevier, San Diego, California, pp. 157–174.

Dudley, R., DeVries, P., 1990. Tropical rain forest structure and the geographical distribution of gliding vertebrates. Biotropica 22, 432–434.

Dudley, R., Ellington, C.P., 1990. Mechanics of forward flight in bumblebees. II. Quasi-steady lift and power requirements. Journal of Experimental Biology 148, 53–58.

Dudley, R., King, V.A., Wassersug, R.J., 1991. The implications of shape and metamorphosis for drag forces on a generalized pond tadpole (*Rana catesbeiana*). Copeia 1991, 252–257.

Dudley, R., Winter, Y., 2002. Hovering flight mechanics of neotropical flower bats (Phyllostomidae : Glossophaginae) in normodense and hypodense gas mixtures. Journal of Experimental Biology 205, 3669–3677.

Eldredge, J.D., 2006. Numerical simulations of undulatory swimming at moderate Reynolds number. Bioinspiration and Biomimetics 1, S19–S24.

Ellington, C.P., 1984a. The aerodynamics of flapping animal flight. American Zoologist 24, 95–105.

Ellington, C.P., 1984b. The aerodynamics of hovering insect flight. I. The quasisteady analysis. Philosophical Transactions of the Royal Society of London B: Biological Sciences 305, 1–15.

Ellington, C.P., 1984c. The aerodynamics of hovering insect flight. IV. Aerodynamic mechanisms. Philosophical Transactions of the Royal Society of London B: Biological Sciences 305, 79–113.

Ellington, C.P., Van den Berg, C., Willmot, A.P., Thomas, A.L.R., 1996. Leading-edge vortices in insect flight. Nature 384, 626–630.

Elzinga, M.J., van Breugel, F., Dickinson, M.H., 2014. Strategies for the stabilization of longitudinal forward flapping flight revealed using a dynamically-scaled robotic fly. Bioinspiration and Biomimetics 9, 025001. http://dx.doi.org/10.1088/1748-3182/9/2/025001.

Ennos, A.R., 1988. The importance of torsion in the design of insect wings. Journal of Experimental Biology 140, 137–160.

Ennos, A.R., 1989. Inertial and aerodynamic torques on the wings of Diptera in flight. Journal of Experimental Biology 142, 87–95.

Ennos, A.R., 1993. The mechanics of the flower stem of the sedge *Carex acutiformis*. Annals of Botany 72, 123–127.

Ennos, A.R., 2012. Solid Biomechanics. Princeton University Press, Princeton, New Jersey. 264 pp.

Ennos, A.R., Spatz, H.C., Speck, T., 2000. The functional morphology of the petioles of the banana, *Musa textilis*. Journal of Experimental Botany 51, 2085–2093.

Ennos, A.R., van Casteren, A., 2010. Transverse stresses and modes of failure in tree branches and other beams. Proceedings of the Royal Society of London B: Biological Sciences 277, 1253–1258.

Espino, D.M., Shepherd, D.E.T., Hukins, D.W.L., 2014. Viscoelastic properties of bovine knee joint articular cartilage: dependency on thickness and loading frequency. BMC Musculo-skeletal Disorders 15, 2–23.

Etnier, S.A., Vogel, S., 2000. Reorientation of daffodil (*Narcissus*: Amaryllidaceae) flowers in wind: drag reduction and torsional flexibility. American Journal of Botany 87, 29–32.

Farran, L., Ennos, A.R., Eichhorn, S.J., 2008. The effect of humidity on the fracture properties of human fingernails. Journal of Experimental Biology 211, 3677–3681.

Feldkamp, S.D., 1987. Swimming in the California sea lion: morphometrics, drag and energetics. Journal of Experimental Biology 131, 117–135.

Felsenstein, J., 1985. Phylogenies and the comparative method. The American Naturalist 125, 1–15.

Fox, R.W., McDonald, A.T., 1998. Introduction to Fluid Mechanics, fifth ed. John Wiley & Sons, Inc., New York. 762 pp.

Frost, H.M., 1967. An Introduction to Biomechanics. C.C. Thomas, Springfield, Illinois, 151 pp.

Full, R.J., Tu, M.S., 1991. Mechanics of a rapid running insect: two-, four- and six-legged loco-motion. Journal of Experimental Biology 156, 215–231.

Fung, Y.-C., 1981. Biomechanics: Mechanical Properties of Living Tissues. Springer-Verlag, New York, 433 pp.

Gerdes, J.W., Wilkerson, S.A., Gupta, S.K., 2012. A review of bird-inspired flapping wing miniature air vehicle designs. Journal of Mechanisms and Robotics 4, 021003. http://dx.doi.org/10.1115/1.4005525.

Gibbs, D.A., Merril, E.W., Smith, K.A., Balazs, E.A., 1968. Rheology of hyaluronic acid. Biopolymers 6, 777–791.

Goldspink, G., 1980. Locomotion and the sliding filament mechanism. In: Elder, H.Y., Trueman, E.R. (Eds.), Aspects of Animal Locomotion. Cambridge University Press, Cambridge, UK, pp. 1–25.

Gordon, J.E., 1976. The New Science of Strong Materials, second ed. Princeton University Press, Princeton, New Jersey. 287 pp.

Gordon, J.E., 1978. Structures, or Why Things Don't Fall Down. Penguin Books, Harmondsworth, Middlesex, England, 395 pp.

Gosline, J.M., 1971a. Connective tissue mechanics of *Metridium senile*. I. Structural and compositional aspects. Journal of Experimental Biology 55, 763–774.

Gosline, J.M., 1971b. Connective tissue mechanics of *Metridium senile*. II. Visco-elastic properties and macromolecular model. Journal of Experimental Biology 55, 775–795.

Grace, J., 1977. Plant Response to Wind. Academic Press, New York, 204 pp.

Grace, J., Wilson, J., 1976. The boundary layer over a *Populus* leaf. Journal of Experimental Botany 27, 231–241.

Gray, J., Hancock, G.J., 1955. The propulsion of sea-urchin spermatozoa. Journal of Experimental Biology 32, 802–814.

Hainsworth, F.R., 1989. Wing movements and positioning for aerodynamic benefit by Canada geese flying in formation. Canadian Journal of Zoology 67, 585–589.

Haldane, J.B.S., 1927. Possible Worlds, and Other Essays. Chatto & Windus, London, 312 pp.

Halliday, D., Resnick, R., Walker, J., 2010. Fundamentals of Physics, ninth ed. John Wiley & Sons, New York. 1136 pp.

Hamdani, H.R., Naqvi, A., 2011. A study on the mechanism of high-lift generation by an insect wing in unsteady motion at small Reynolds number. International Journal for Numerical Methods in Fluids 67, 581–598.

Han, J.S., Chang, J.W., Cho, H.K., 2015. Vortices behavior depending on the aspect ratio of an insect-like flapping wing in hover. Experiments in Fluids 56, 181. http://dx.doi.org/10.1007/s00348-015-2049-9.

Harkness, M.L.R., Harkness, R.D., 1959. Changes in the physical properties of the uterine cervix of the rat during pregnancy. Journal of Physiology 148, 524–547.

Harper, D.G., Blake, R.W., 1990. Fast-start performance of rainbow trout *Salmo gairdneri* and northern pike *Esox lucius*. Journal of Experimental Biology 150, 321–342.

Haut, R.C., Little, R.W., 1969. Rheological properties of canine anterior cruciate ligaments. Journal of Biomechanics 2, 289–292.

Herbert, R.C., Young, P.G., Smith, C.W., Wootton, R.J., Evans, K.E., 2000. The hind wing of the desert locust (*Schistocerca gregaria* Forskal). III. A finite element analysis of a deployable structure. Journal of Experimental Biology 203, 2945–2955.

Herzog, W. (Ed.), 2000. Skeletal Muscle Mechanics: From Mechanisms to Function. John Wiley & Sons, New York, 554 pp.

Higuchi, T., Nagami, T., Nakata, H., Watanabe, M., Isaka, T., Kanosue, K., 2016. Contribution of visual information about ball trajectory to baseball hitting accuracy. PLoS One 11, e0148498. http://dx.doi.org/10.1371/journal.pone.0148498.

Hill, A.V., 1950. The dimensions of animals and their muscular dynamics. Science Progress 38, 209–230.

Hill, R.W., Wyse, G.A., Anderson, M., 2016. Animal Physiology, fourth ed. Sinauer Associates, Sunderland, Mass. 828 pp.

Hoerner, S.F., 1965. Fluid-Dynamic Drag. Hoerner Fluid Dynamics, Bricktown, New Jersey, 438 pp.

Hoyt, D.F., Taylor, C.R., 1981. Gait and the energetics of locomotion in horses. Nature 292, 239–240.

Hu, D.L., Sielert, K., Gordon, M., 2011. Turtle shell and mammal skull resistance to fracture due to predator bites and ground impact. Journal of Mechanics of Materials and Structures 6, 1197–1211.

Ishihara, D., Yamashita, Y., Horie, T., Yoshida, S., Niho, T., 2009. Passive maintenance of high angle of attack and its lift generation during flapping translation in crane fly wing. Journal of Experimental Biology 212, 3882–3891.

Isnard, S., Cobb, A.R., Holbrook, N.M., Zwieniecki, M., Dumais, J., 2009. Tensioning the helix: a mechanism for force generation in twining plants. Proceedings of the Royal Society B: Biological Sciences 276, 2643–2650.

Jensen, M., 1956. Biology and physics of locust flight. III. The aerodynamics of locust flight. Philosophical Transactions of the Royal Society of London B: Biological Sciences 239, 511–552.

Jeronimidis, G., 1980. Wood, one of nature's challenging composites. In: Vincent, J.F.V., Currey, J.D. (Eds.), The Mechanical Properties of Biological Materials, vol. 34. Cambridge University Press, Cambridge, UK, pp. 169–182.

Jørgensen, C.B., 1955. Quantitative aspects of filter feeding in invertebrates. Biological Reviews of the Cambridge Philosophical Society 30, 391–454.

Jung, S., 2010. *Caenorhabditis elegans* swimming in a saturated particulate system. Physics of Fluids 22, 1–6.

Kapsali, V., 2016. Biomimicry for Designers: Applying Nature's Processes and Materials in the Real World. Thames & Hudson, London, 240 pp.

Kasapi, M.A., Gosline, J.M., 1996. Strain-rate-dependent mechanical properties of the equine hoof wall. Journal of Experimental Biology 199, 1133–1146.

Kelly, D.A., 2007. Penises as variable-volume hydrostatic skeletons. Annals of the New York Academy of Sciences 1101, 453–463.

Ker, R.F., Bennett, M.B., Bibby, S.R., Kester, R.C., Alexander, R.M., 1987. The spring in the arch of the human foot. Nature 325, 147–149.

Kier, W.M., 1985. The musculature of squid arms and tentacles: ultrastructural evidence for functional differences. Journal of Morphology 185, 223–239.

Kier, W.M., 2012. The diversity of hydrostatic skeletons. Journal of Experimental Biology 215, 1247–1257.

Kier, W.M., Smith, K.K., 1985. Tongues, tentacles and trunks: the biomechanics of movement in muscular-hydrostats. Zoological Journal of the Linnean Society 83, 307–324.

Koehl, M.A.R., 1995. Fluid flow through hair-bearing appendages: feeding, smelling, and swimming at low and intermediate Reynolds numbers. In: Ellington, C.P., Pedley, T.J. (Eds.), Biological Fluid Dynamics, vol. 49. Company of Biologists Ltd, Cambridge, UK, pp. 157–182.

Kohannim, S., Iwasaki, T., 2014. Analytical insights into optimality and resonance in fish swimming. Journal of the Royal Society Interface 11, 20131073. http://dx.doi.org/10.1098/rsif.2013.1073.

Kolomenskiy, D., Moffatt, H.K., Farge, M., Schneider, K., 2011. The Lighthill-Weis-Fogh clap-fling-sweep mechanism revisited. Journal of Fluid Mechanics 676, 572−606.

LaBarbera, M., 1981. Water-flow patterns in and around 3 species of articulate brachiopods. Journal of Experimental Marine Biology and Ecology 55, 185−206.

Labarbera, M., 1983. Why the wheels won't go. American Naturalist 121, 395−408.

Lai, J.H., del Alamo, J.C., Rodriguez-Rodriguez, J., Lasheras, J.C., 2010. The mechanics of the adhesive locomotion of terrestrial gastropods. Journal of Experimental Biology 213, 3920−3933.

Lauder, G.V., 2015. Fish locomotion: recent advances and new directions. Annual Review of Marine Science 7, 521−545.

Lauder, G.V., Tytell, E.D., 2006. Hydrodynamics of undulatory propulsion. In: Shadwick, R.E., Lauder, G.V. (Eds.), Fish Biomechanics. Academic Press (Elsevier), San Diego, California, pp. 425−468.

Lewis, A.M., 1992. Measuring the hydraulic diameter of a pore or conduit. American Journal of Botany 79, 1158−1161.

Li, H.D., Ang, H.S., 2016. Preliminary airfoil design of an innovative adaptive variable camber compliant wing. Journal of Vibroengineering 18, 1861−1873.

Lighthill, M.J., 1971. Large-amplitude elongated-body theory of fish locomotion. Proceedings of the Royal Society B: Biological Sciences 179, 125−138.

Lighthill, M.J., 1973. On the Weis-Fogh mechanism of lift generation. Journal of Fluid Mechanics 60, 1−17.

Lighthill, M.J., 1975. Aerodynamic aspects of animal flight. In: Wu, T.Y.-T., Brokaw, C.J., Brennan, C. (Eds.), Swimming and Flying in Nature, vol. 2. Plenum Press, New York, pp. 423−491.

Liu, H., Ravi, S., Kolomenskiy, D., Tanaka, H., 2016. Biomechanics and biomimetics in insect-inspired flight systems. Philosophical Transactions of the Royal Society B: Biological Sciences 371, 20150390. http://dx.doi.org/10.1098/rstb.2015.0390.

Liu, Y.P., Sun, M., 2008. Wing kinematics measurement and aerodynamics of hovering droneflies. Journal of Experimental Biology 211, 2014−2025.

Lohse, D., Rauhe, R., Bergmann, R., van der Meer, D., 2004. Granular physics: creating a dry variety of quicksand. Nature 432, 689−690.

Lomakin, J., Huber, P.A., Eichler, C., Arakane, Y., Kramer, K.J., Beernan, R.W., Kanost, M.R., Gehrke, S.H., 2011. Mechanical properties of the beetle elytron, a biological composite material. Biomacromolecules 12, 321−335.

Long Jr., J.H., Schumacher, J., Livingston, N., Kemp, M., 2006. Four flippers or two? Tetrapodal swimming with an aquatic robot. Bioinspiration and Biomimetics 1, 20−29.

Maeng, J.S., Park, J.H., Jang, S.M., Han, S.Y., 2013. A modeling approach to energy savings of flying Canada geese using computational fluid dynamics. Journal of Theoretical Biology 320, 76−85.

Maladen, R.D., Ding, Y., Li, C., Goldman, D.I., 2009. Undulatory swimming in sand: subsurface locomotion of the sandfish lizard. Science 325, 314−318.

Maladen, R.D., Ding, Y., Umbanhowar, P.B., Kamor, A., Goldman, D.I., 2011. Mechanical models of sandfish locomotion reveal principles of high performance subsurface sand-swimming. Journal of the Royal Society Interface 8, 1332−1345.

Manning, P.L., Margetts, L., Johnson, M.R., Withers, P.J., Sellers, W.I., Falkingham, P.L., Mummery, P.M., Barrett, P.M., Raymont, D.R., 2009. Biomechanics of dromaeosaurid dinosaur claws: application of x-ray microtomography, nanoindentation, and finite element analysis. The Anatomical Record: Advances in Integrative Anatomy and Evolutionary Biology 292, 1397−1405.

Marsh, R.L., 1994. Jumping ability of anuran amphibians. Advances in Veterinary Science and Comparative Medicine 38B, 51−111.

McGreer, T., 1992. Principles of walking and running. In: Alexander, R.M. (Ed.), Mechanics of Animal Locomotion. Springer-Verlag, New York, pp. 114−139.

McMahon, T.A., 1984. Muscles, Reflexes, and Locomotion. Princeton University Press, Princeton, New Jersey, 331 pp.

Minotti, F.O., 2002. Unsteady two-dimensional theory of a flapping wing. Physical Review E 66, 10.

Mish, F.C. (Ed.), 1983. Webster's Ninth New Collegiate Dictionary. Merriam-Webster, Inc, Springfield, Massachusetts.

Monn, M.A., Weaver, J.C., Zhang, T.Y., Aizenberg, J., Kesari, H., 2015. New functional insights into the internal architecture of the laminated anchor spicules of *Euplectella aspergillum*. Proceedings of the National Academy of Sciences of the United States of America 112, 4976−4981.

Mou, X.L., Liu, Y.P., Sun, M., 2011. Wing motion measurement and aerodynamics of hovering true hoverflies. Journal of Experimental Biology 214, 2832−2844.

Nachtigall, W., 1980. Mechanics of swimming in water-beetles. In: Elder, H.Y., Trueman, E.R. (Eds.), Aspects of Animal Movement. Cambridge University Press, Cambridge, UK, pp. 107−124.

Nelson, E.W., Best, C., McLean, W.G., 1997. Schaum's Outline of Engineering Mechanics: Statics and Dynamics, fifth ed. McGraw-Hill, New York. 480 pp.

Niklas, K.J., 1992. Plant Biomechanics: An Engineering Approach to Plant Form and Function. University of Chicago Press, Chicago, 607 pp.

Norberg, U.M., 1975. Hovering flight in the pied flycatcher (*Ficedula hypoleuca*). In: Wu, T.Y.-T., Brokaw, C.J., Brennan, C. (Eds.), Swimming and Flying in Nature, vol. 2. Plenum Press, New York, pp. 869−881.

Norberg, U.M., 1976a. Aerodynamics, kinematics, and energetics of horizontal flapping flight in the long-eared bat *Plecotus auritus*. Journal of Experimental Biology 65, 179−212.

Norberg, U.M., 1976b. Aerodynamics of hovering flight in long-eared bat *Plecotus auritus*. Journal of Experimental Biology 65, 459−470.

Nudds, R.L., Taylor, G.K., Thomas, A.L.R., 2004. Tuning of Strouhal number for high propulsive efficiency accurately predicts how wingbeat frequency and stroke amplitude relate and scale with size and flight speed in birds. Proceedings of the Royal Society of London B: Biological Sciences 271, 2071−2076.

Ogston, A.G., Stanier, J.E., 1953. The physiological function of hyaluronic acid in synovial fluid; viscous, elastic and lubricant properties. The Journal of Physiology 119, 244−252.

Osborne, M.F.M., 1951. Aerodynamics of flapping flight with application to insects. Journal of Experimental Biology 28, 221−245.

Parle, E., Herbaj, S., Sheils, F., Larmon, H., Taylor, D., 2016. Buckling failures in insect exoskeletons. Bioinspiration and Biomimetics 11, 016003. http://dx.doi.org/10.1088/1748-3190/11/1/016003.

Pennycuick, C.J., 1968. A wind-tunnel study of gliding flight in the pigeon *Columba livia*. Journal of Experimental Biology 49, 509−526.

Pennycuick, C.J., 1971. Gliding flight of the white-backed vulture *Gyps africanus*. Journal of Experimental Biology 55, 13—38.

Pennycuick, C.J., 1972. Animal Flight. Edward Arnold, London, 68 pp.

Pennycuick, C.J., Obrecht, H.H., Fuller, M.R., 1988. Empirical estimates of body drag of large waterfowl and raptors. Journal of Experimental Biology 135, 253—264.

Pérez-Rigueiro, J., Viney, C., Llorca, J., Elices, M., 1998. Silkworm silk as an engineering material. Journal of Applied Polymer Science 70, 2439—2447.

Placet, V., Passard, J., Perre, P., 2007. Viscoelastic properties of green wood across the grain measured by harmonic tests in the range 0—95 degrees C: hardwood vs. softwood and normal wood vs. reaction wood. Holzforschung 61, 548—557.

Preetha, A., Banerjee, R., 2005. Comparison of artificial saliva substitutes. Trends in Biomaterials and Artificial Organs 18, 178—186.

Previtali, F., Arrieta, A.F., Ermanni, P., 2016. Investigation of the optimal elastic and weight properties of passive morphing skins for camber-morphing applications. Smart Materials and Structures 25, 055040. http://dx.doi.org/10.1088/0964-1726/25/5/055040.

Raibert, M., Blankespoor, K., Nelson, G., Playter, R., 2008. BigDog, the Rough-Terrain Quaduped Robot. Boston Dynamics. http://www.bostondynamics.com/img/BigDog_IFAC_Apr-8-2008.pdf.

Reis, P.M., Jung, S.H., Aristoff, J.M., Stocker, R., 2010. How cats lap: water uptake by *Felis catus*. Science 330, 1231—1234.

Reynolds, O., 1883. An experimental investigation of the circumstances which determine whether the motion of water shall be direct or sinuous, and the laws of resistance in parallel channels. Philosophical Transactions of the Royal Society of London 174, 935—982.

Reynolds, S.E., 1975. The mechanical properties of the abdominal cuticle of *Rhodnius* larvae. Journal of Experimental Biology 62, 69—80.

Rigby, B.J., Hirai, N., Spikes, J.D., Eyring, H., 1959. The mechanical properties of rat tail tendon. Journal of General Physiology 43, 265—283.

Rodenborn, B., Chen, C.H., Swinney, H.L., Liu, B., Zhang, H.P., 2013. Propulsion of microorganisms by a helical flagellum. Proceedings of the National Academy of Sciences of the United States of America 110, E338—E347.

Ronken, S., Arnold, M.P., Garcia, H.A., Jeger, A., Daniels, A.U., Wirz, D., 2012. A comparison of healthy human and swine articular cartilage dynamic indentation mechanics. Biomechanics and Modeling in Mechanobiology 11, 631—639.

Sane, S.P., 2003. The aerodynamics of insect flight. Journal of Experimental Biology 206, 4191—4208.

Sane, S.P., Dickinson, M.H., 2002. The aerodynamic effects of wing rotation and a revised quasi-steady model of flapping flight. Journal of Experimental Biology 205, 1087—1096.

Santhanakrishnan, A., Dollinger, M., Hamlet, C.L., Colin, S.P., Miller, L.A., 2012. Flow structure and transport characteristics of feeding and exchange currents generated by upside-down *Cassiopea* jellyfish. Journal of Experimental Biology 215, 2369—2381.

Schmidt-Nielsen, K., 1984. Scaling: Why Is Animal Size so Important? Cambridge University Press, Cambridge, UK, 241 pp.

Secomb, T.W., 1995. Mechanics of blood flow in the microcirculation. In: Ellington, C.P., Pedley, T.J. (Eds.), Biological Fluid Dynamics. The Company of Biologists Ltd, Cambridge, UK, pp. 305—321.

Seebacher, F., 2003. Dinosaur body temperatures: the occurrence of endothermy and ectothermy. Paleobiology 29, 105—122.

Shadwick, R.E., 1999. Mechanical design in arteries. Journal of Experimental Biology 202, 3305—3313.

Shadwick, R.E., Gosline, J.M., 1985. Mechanical properties of the octopus aorta. Journal of Experimental Biology 114, 259–284.

Shadwick, R.E., Gosline, J.M., 1995. Arterial windkessels in marine mammals. In: Ellington, C.P., Pedley, T.J. (Eds.), Biological Fluid Dynamics. The Company of Biologists Ltd, Cambridge, UK, pp. 243–252.

Shaughnessy Jr., E.J., Katz, I.M., Schaffer, J.P., 2005. Introduction to Fluid Mechanics. Oxford University Press, New York, 1042 pp.

Shen, C., Sun, M., 2015. Power requirements of vertical flight in the dronefly. Journal of Bionic Engineering 12, 227–237.

Shephard, K.L., 1994. Functions for fish mucus. Reviews in Fish Biology and Fisheries 4, 401–429.

Sherwood, L., Klandorf, H., Yancey, P., 2013. Animal Physiology: From Genes to Organisms, second ed. Brooks/Cole, Belmont, California. 896 pp.

Shyy, W., Kang, C.K., Chirarattananon, P., Ravi, S., Liu, H., 2016. Aerodynamics, sensing and control of insect-scale flapping-wing flight. Proceedings of the Royal Society A: Mathematical Physical and Engineering Sciences 472, 20150712. http://dx.doi.org/10.1098/rspa.2015.0712.

Speakman, J.R., Banks, D., 1998. The function of flight formations in Greylag geese *Anser anser*; energy saving or orientation? IBIS 140, 280–287.

Spedding, G.R., 1986. The wake of a jackdaw (*Corvus monedula*) in slow flight. Journal of Experimental Biology 125, 287–307.

Sun, M., Tang, H., 2002. Unsteady aerodynamic force generation by a model fruit fly wing in flapping motion. Journal of Experimental Biology 205, 55–70.

Sun, M., Xiong, Y., 2005. Dynamic flight stability of a hovering bumblebee. Journal of Experimental Biology 208, 447–459.

Sutton, G.P., Burrows, M., 2011. Biomechanics of jumping in the flea. Journal of Experimental Biology 214, 836–847.

Swartz, S.M., Bennett, M.B., Carrier, D.R., 1992. Wing bone stresses in free flying bats and the evolution of skeletal design for flight. Nature 359, 726–729.

Taylor, G.M., 2000. Maximum force production: why are crabs so strong? Proceedings of the Royal Society B: Biological Sciences 267, 1475–1480.

Tedrake, R., Zhang, T.W., Fong, M-F., Seung, H.S., 2004. Actuating a simple 3D passive dynamic walker. In: Robotics and Automation, 2004. Proceedings. ICRA'04. 2004 IEEE International Conference on. IEEE. http://dx.doi.org/10.1109/ROBOT.2004.1302452.

Tennekes, H., 1996. The Simple Science of Flight: From Insects to Jumbo Jets. MIT Press, Cambridge, Massachusetts, 137 pp.

Tennekes, H., 2009. The Simple Science of Flight: From Insects to Jumbo Jets (revised and expanded ed.). MIT Press, Cambridge, Massachusetts, 201 pp.

Thom, A., Swart, P., 1940. The forces on an aerofoil at very low speeds. Journal of the Royal Aeronautical Society 44, 761–770.

Thorington Jr., R.W., Heaney, L.R., 1981. Body proportions and gliding adaptations of flying squirrels (Petauristinae). Journal of Mammalogy 62, 101–114.

Torricelli, D., Gonzalez, J., Weckx, M., Jiménez-Fabián, R., Vanderborght, B., Sartori, M., Dosen, S., Farina, D., Lefeber, D., Pons, J.L., 2016. Human-like compliant locomotion: state of the art of robotic implementations. Bioinspiration and Biomimetics 11, 051002. http://dx.doi.org/10.1088/1748-3190/11/5/051002.

Trueman, E.R., 1953. Observations on certain mechanical properties of the ligament of *Pecten*. Journal of Experimental Biology 30, 453–467.

Tucker, V.A., 1990a. Body drag, feather drag and interference drag of the mounting strut in a peregrine falcon, *Falco peregrinus*. Journal of Experimental Biology 149, 449—468.

Tucker, V.A., 1990b. Measuring aerodynamic interference drag between a bird body and the mounting strut of a drag balance. Journal of Experimental Biology 154, 439—461.

Tucker, V.A., 1993. Gliding birds: reduction of induced drag by wing tip slots between the primary feathers. Journal of Experimental Biology 180, 285—310.

Tucker, V.A., 1995. Drag reduction by wing tip slots in a gliding Harris' hawk, *Parabuteo unicinctus*. Journal of Experimental Biology 198, 775—781.

Tucker, V.A., 2000. Gliding flight: drag and torque of a hawk and a falcon with straight and turned heads, and a lower value for the parasite drag coefficient. Journal of Experimental Biology 203, 3733—3744.

Tucker, V.A., Heine, C., 1990. Aerodynamics of gliding flight in a Harris' hawk *Parabuteo unicinctus*. Journal of Experimental Biology 149, 469—490.

Tucker, V.A., Parrott, G.C., 1970. Aerodynamics of gliding flight in a falcon and other birds. Journal of Experimental Biology 52, 345—367.

Van den Berg, C., Ellington, C.P., 1997a. The vortex wake of a "hovering" model hawkmoth. Philosophical Transactions of the Royal Society of London B: Biological Sciences 352, 317—328.

Van den Berg, C., Ellington, C.P., 1997b. The three-dimensional leading edge vortex of a "hovering" model hawkmoth. Philosophical Transactions of the Royal Society of London B: Biological Sciences 352, 329—340.

van Leeuwen, J.L., 1992. Muscle function in locomotion. In: Alexander, R.M. (Ed.), Mechanics of Animal Locomotion. Springer-Verlag, New York, pp. 191—250.

Ventre, M., Mollica, F., Netti, P.A., 2009. The effect of composition and microstructure on the viscoelastic properties of dermis. Journal of Biomechanics 42, 430—435.

Versluis, M., Blom, C., van der Meer, D., van der Weele, K., Lohse, D., 2006. Leaping shampoo and the stable Kaye effect. Journal of Statistical Mechanics: Theory and Experiment, P07007. http://dx.doi.org/10.1088/1742-5468/2006/07/p07007.

Videler, J.J., Stamhuis, E.J., Povel, G.D.E., 2004. Leading-edge vortex lifts swifts. Science 306, 1960—1962.

Vincent, J.F.V., 1982. Structural Biomaterials. Macmillan, London, 206 pp.

Vincent, J.F.V., 1990. Structural Biomaterials. Princeton University Press, Princeton, New Jersey, 206 pp.

Vincent, J.F.V., Wood, S.D.E., 1972. Mechanism of abdominal extension during oviposition in *Locusta*. Nature 235, 167—168.

Vogel, S., 1966. Flight in *Drosophila*. I. Flight performance of tethered flies. Journal of Experimental Biology 44, 567—578.

Vogel, S., 1967. Flight in *Drosophila*. III. Aerodynamic characteristics of fly wings and wing models. Journal of Experimental Biology 46, 431—443.

Vogel, S., 1981. Life in Moving Fluids: The Physical Biology of Flow, first ed. Willard Grant Press, Boston. 352 pp.

Vogel, S., 1988. Life's Devices: The Physical World of Animals and Plants. Princeton University Press, Princeton, New Jersey, 367 pp.

Vogel, S., 1992. Twist-to-bend ratios and cross-sectional shapes of petioles and stems. Journal of Experimental Botany 43, 1527—1532.

Vogel, S., 1994. Life in Moving Fluids: The Physical Biology of Flow, second ed. Princeton University Press, Princeton, New Jersey. 467 pp.

Vogel, S., 1995. Twist-to-bend ratios of woody structures. Journal of Experimental Botany 46, 981–985.

Vogel, S., 1998. Cats' Paws and Catapults: Mechanical Worlds of Nature and People. W.W. Norton & Co, New York, 382 pp.

Vogel, S., 2001. Prime Mover: A Natural History of Muscle. W.W. Norton & Co, New York, 370 pp.

Vogel, S., 2003. Comparative Biomechanics: Life's Physical World. Princeton University Press, Princeton, New Jersey, 580 pp.

Vogel, S., 2009. Glimpses of Creatures in Their Physical Worlds. Princeton University Press, Princeton, New Jersey, 320 pp.

Vogel, S., 2013. Comparative Biomechanics: Life's Physical World, second ed. Princeton University Press, Princeton, New Jersey. 628 pp.

Vogel, S., Papanicolaou, M.N., 1983. A constant stress creep testing machine. Journal of Biomechanics 16, 153–156.

von Busse, R., Waldman, R.M., Swartz, S.M., Voigt, C.C., Breuer, K.S., 2014. The aerodynamic cost of flight in the short-tailed fruit bat (*Carollia perspicillata*): comparing theory with measurement. Journal of the Royal Society Interface 11, 20140147. http://dx.doi.org/10.1098/rsif.2014.0147.

Vosburgh, F., 1982. *Acropora reticulata*: structure, mechanics and ecology of a reef coral. Proceedings of the Royal Society of London B: Biological Sciences 214, 481–499.

Wainwright, S.A., Biggs, W.D., Curry, J.D., Gosline, J.M., 1976. Mechanical Design in Organisms. John Wiley & Sons, New York, 423 pp.

Wainwright, S.A., Biggs, W.D., Curry, J.D., Gosline, J.M., 1982. Mechanical Design in Organisms (paperback ed.). Princeton University Press, Princeton, New Jersey, 423 pp.

Ward, D.V., Wainwright, S.A., 1972. Locomotory aspects of squid mantle structure. Journal of Zoology 167, 437–449.

Wardle, C.S., Videler, J.J., 1980. Fish swimming. In: Elder, H.Y., Trueman, E.R. (Eds.), Aspects of Animal Movement. Cambridge University Press, Cambridge, UK, pp. 125–150.

Washburn, J.O., Washburn, L., 1984. Active aerial dispersal of minute wingless arthropods: exploitation of boundary-layer velocity gradients. Science 223, 1088–1089.

Webb, P.W., 1975. Hydrodynamics and energetics of fish propulsion. Bulletin of the Fisheries Research Board of Canada 190, 1–158.

Webb, P.W., 1976. The effect of size on the fast-start performance of rainbow trout *Salmo gairdneri*, and a consideration of piscivorous predator-prey interactions. Journal of Experimental Biology 65, 157–177.

Webb, P.W., 1979. Mechanics of escape responses in crayfish (*Orconectes virilis*). Journal of Experimental Biology 79, 245–263.

Weis-Fogh, T., 1956. Biology and physics of locust flight. II. Flight performance of the desert locust (*Schistocerca gregaria*). Philosophical Transactions of the Royal Society of London B: Biological Sciences 239, 459–510.

Weis-Fogh, T., 1960. A rubber-like protein in insect cuticle. Journal of Experimental Biology 37, 889–907.

Weis-Fogh, T., 1973. Quick estimates of flight fitness in hovering animals, including novel mechanisms for lift production. Journal of Experimental Biology 59, 169–230.

Weis-Fogh, T., 1975. Flapping flight and power in birds and insects, conventional and novel mechanisms. In: Wu, T.Y.-T., Brokaw, C.J., Brennan, C. (Eds.), Swimming and Flying in Nature, vol. 2. Plenum Press, New York, pp. 729–762.

Weis-Fogh, T., Jensen, M., 1956. Biology and physics of locust flight. I. Basic principles of insect flight: a critical review. Philosophical Transactions of the Royal Society of London B: Biological Sciences 239, 415–457.

Wilkie, D.R., 1960. Man as a source of mechanical power. Ergonomics 3, 1–8.

Withers, P.C., 1992. Comparative Animal Physiology. Saunders College Publishing, Fort Worth, Texas, 949 pp.

Woesz, A., Weaver, J.C., Kazanci, M., Dauphin, Y., Aizenberg, J., Morse, D.E., Fratzl, P., 2006. Micromechanical properties of biological silica in skeletons of deep-sea sponges. Journal of Materials Research 21, 2068–2078.

Wood, R.J., Avadhanula, S., Sahai, R., Steltz, E., Fearing, R.S., 2008. Microrobot design using fiber reinforced composites. Journal of Mechanical Design 130, 052304. http://dx.doi.org/10.1115/1.2885509.

Wu, T.Y.-T., 1971. Hydromechanics of swimming propulsion. 1. Swimming of 2-dimensional flexible plate at variable forward speeds in an inviscid fluid. Journal of Fluid Mechanics 46, 337–355.

Yamamoto, M., Isogai, K., 2005. Direct measurement of unsteady fluid dynamic forces for a hovering dragonfly. AIAA Journal 43, 2475–2480.

Yildiz, M.Z., Güçül, B., 2013. Relationship between vibrotactile detection threshold in the Pacinian channel and complex mechanical modulus of the human glabrous skin. Somatosensory and Motor Research 30, 37–47.

Zarek, J.M., 1959. Biomechanics. Nature 184, 512–513.

Zaret, R.E., Kerfoot, W.C., 1980. Shape and swimming technique of *Bosmina longirostris*. Limnology and Oceanography 25, 126–133.

Zaslansky, P., Shahar, R., Friesem, A.A., Weiner, S., 2006. Relations between shape, materials properties, and function in biological materials using laser speckle interferometry: in situ tooth deformation. Advanced Functional Materials 16, 1925–1936.

Żbikowski, R., 2002. On aerodynamic modelling of an insect–like flapping wing in hover for micro air vehicles. Philosophical Transactions of the Royal Society of London. Series A: Mathematical, Physical and Engineering Sciences 360, 273–290.

Żbikowski, R., Ansari, S.A., Knowles, K., 2006. On mathematical modelling of insect flight dynamics in the context of micro air vehicles. Bioinspiration and Biomimetics 1, R26–R37.

Zhang, T.N., Goldman, D.I., 2014. The effectiveness of resistive force theory in granular locomotion. Physics of Fluids 26, 1–17.

Index

'*Note*: Page numbers followed by "f" indicate figures, "t" indicate tables, and "b" indicate boxes.'

A

Acceleration reaction concept, 93–94
Added mass coefficient, 93–94
Aerodynamic efficiency, 76
Aerodynamic models, 156
Alder, G.M., 143
Alexander, David E., 143
Alexander, R. McNeill, 4, 49
Aluminum alloys, 17, 21–23, 27–28
Amorphous rubber elasticity mechanism, 25b
Aneurisms, 133
Anguilliform, 87–88, 88f
Animal circulatory systems, 64
Animal flight analysis, 9
Anisotropic materials, 28, 45
Applied biomechanics, 2
Ariolimax columbianus, 116–117
Arteries
    artery elasticity smoothes blood flow, 105,
        106f
    cephalopods, 105
    dog aortas, 106
    elastin, 105
    vertebrates, 105
    viscous component, 106
Articulation, rigid skeletons
    aquatic arthropods, 124
    arthrodial (flexible) cuticle, 124–125
    arthropods, 124
    ball and socket joints, 125–126
    bone shaft, 125
    buckling, 124
    cartilage, 125
    cord-/straplike ligaments, 126
    degrees of freedom, 124–126
    dicondylic joints, 124–125, 125f
    exoskeleton, 126
    frictional forces, 125
    friction coefficients, 125
    gravitational loads, 124–125
    hinge joints, 125–126
    hip and shoulder joints, 125–126
    joint lubrication, 125
    pegs and sockets, 125
    radius of curvature, 125
    synovial fluid, 125
    terrestrial vertebrates, 124–125
    vertebrate joints, 126
    vertebrate limb joints, 125
Artificial polymers, 101–102
Artificial rubbers, 25b
Aspect ratio, 75–76, 76f
Athletic performance, 2
Autocorrelation/signal processing algorithms,
    155
Autonomous aerial vehicle, 147

B

Balance, pan, 11–12
Barnacles, 147–148
Bat flight mechanics, 4–5
Beams
    anisotropic and non-Hookean materials, 42
    anisotropic material, 39–40
    beam theory, 40
    bicycle pedal/crank handle, 41–42
    biological beam, 40–41
    cantilevers, 39
    deflection, 39–41, 39f
    elongate animal/plant bodies/body parts,
        39
    fixed beams, 39
    geometric property, 40
    isotropic material, 39–40
    material property, 40, 42
    neutral surface, 39–40
    shape property, 42
    shear modulus, 42
    solid circular cross section and hollow
        cylinder, 40, 40f
    stresses, 39–40
    tensile and compressive stress, 40
    tension and elongation strain, 39–40
    three-point bending test, 40–41

Beams (*Continued*)
  torsional compression, 42
  torsional stiffness, 42
Bearings, 152—153
Beiwener, Andrew A., 4—5
Bending, 47
  buckling, 38
  torsion, 45—46
  compliance, 46—47
  twisting, 45
Bennet-Clark, H.C., 143
Bennett, S. Christopher, 143—144
Bernoulli's equation, 51, 63—64
Bioinspiration and bio-inspired design, 148
Biological ceramics, 152
Biomechanics. *See also specific*
    biomechanics
  beyond standard engineering approaches, 1
  vs. biophysics, 3
  clinical biomechanics. *See* Clinical
    biomechanics
  definition, 1
  engineering approaches and techniques, 1
  macroscopic organisms, 2—3
  nonhuman organisms, 1—3
  sports biomechanics, 2
Biomimicry
  adhesives, 147—148
  bivalve mollusks, 146
  bone tissue remodels, 146
  byssal threads, 146
  cockleburs, 146
  house flies, 146
  legged robots, 148—149
  ornithopters, 147
  paper wasp nests, 146
  specialization and optimization, 146
  thorny hedges, 146
Biophysics, definition, 3
Bipedal locomotion, 148—149
Bird flight mechanics, 3—5, 77—79
Blade element analysis, 117—118
Blood
  axial streaking, 113
  biorheology, 113
  blood vessels, 113
  cell surface protein interactions, 112—113
  charge attraction, 112—113
  dynamic viscosity, 112—113
  physical behavior, 113
  plasma, 112—113
  red blood cells, 112—113

  rheology, 113
  shear thinning, 112—113
  suspension, 113
Bone mechanics, 4—5, 15, 17,
    152. *See also* Skeletons
  loss modulus, 105
  pliant biomaterials, 105
  storage modulus, 105
  vertebrate bone, 105
  Young's modulus, 105
Borrelli, Giovanni, 3, 141
Boundary layers
  barnacles, 67—68
  dynamic soaring, 68
  filter-feeding structures, 67—68
  flow velocity, 66—67
  free-stream speed, 68
  gas exchange surfaces, 67—68
  idealized conditions, 67
  laminar boundary layers, 67—68
  linear gradient, 67
  living organisms, 67
  ocean seabirds, 68
  oxygen-laden water, 68
  potential energy, 68
  Reynolds numbers, 65
  thickness
    fluid density, 65—66
    free-stream fluid speed, 65—66
    laminar boundary layers, 65—66
    leading edge, 65—66
    local Reynolds number, 65—66
    reference length, 65—66
    Reynolds number, 65—66
    surface roughness, 66
    turbulent boundary layer, 66
    viscosity, 65—66
  tiny organisms, 67—68
  turbulent boundary layers, 68
  viscosity, 65
Brown, R.H.J., 3—4
Buckling. *See* Euler buckling, Local buckling

C

Calcite/hydroxyapatite, 37
Carangiform, 87—88, 88f
Caro, C.G., 113
Cartilage, 17
  collagen, 23—24
  cushioning ability, 106—107
  dynamic testing, 106—107

intervertebral disks, 106–107
joints, 106–107
loss modulus, 106–107
three-dimensional meshwork, 106–107
time-dependent component, 106–107
Cauchy strain, 18
Cell membrane process, 3
Cellulose fibers, 45, 144
Chionactis occipitalis, 119
Chitin, 28
Circular/circumferential fibers, 133
Circulatory systems, 52–53
Circumferential stress, 133
"Clap-fling" mechanism, 84, 96
Clinical biomechanics, 1–2
Coefficient of viscosity, 53
Collagen, 23–24, 133–134, 152
  with cartilage, 106–107
  hydrostatic skeletons, 133–134
  with mesoglea, 110–111
  pliant/rigid biomaterials, 28
  ropelike tendons and stiffer cartilage, 37
  structures properties, 100f
  vertebrate joints, 126
Columns
  buckling, 42–43
  compressive stress, 43
  critical force, 43
  deflections/failure loads, 44
  Euler buckling, 42–43, 43f
  flaws and impurities, 44
  Griffith crack propagation, 43
  insect leg segments, 44
  leg bones, 44
  local buckling, 43–44
  plant stems, 44
  sea anemone columns, 44
  squat columns, 42–43
  stress–strain relationship, 42–43
  tensile structures, 42–43
  tree trunks, 44
Comparative biomechanics, 2–3
Compass gait, 127–128
Compliant structures, 155
Compressible flow, 51b
Computational fluid dynamics (CFD)
  techniques, 5, 156
Computer-aided optical correlation, 155
Computer-controlled models, 155–156
Computing-intensive techniques, 96
Condyles, 124–125
Crack growth inhibition

blunt cracks, 37
cartilage, 37
chemical bond strength, 35
compact/cortical bone, 37–38
composite material, 36
crack-blunting voids, 37
cracks and crack propagation reduction, 35,
  36f
critical crack length, 35
ductility, 36
echinoderms, 37
Euplectella aspergillum, 38
fiberglass, 36, 38
fiber-to-matrix interface, 36
firmer matrix, 37
fragile buttons, 37–38
glass, 35
  sponges/hexinactinellids, 38
Gothic cathedral structure, 36
Griffith crack propagation, 35
"holey," mineralized structure, 37
hydroxyapatite crystals, 37–38
material and geometric properties, 35
mild steel and aluminum alloys, 36
mineral crystals, 37–38
mineralized structures, 35
mollusk shells, 37–38
ossicles, 37
reinforced concrete, 36
resin matrix, 36. *See also* Glass; fibers
round/oval voids, 35
sand grain, 38
silica and protein, 38
spicules, 38
spongy bone, 37
tendon, 37
tension-resisting structures, 35
tough isotropic materials, 35
tough protein matrix, 37–38
vertebrate bone tissue, 37–38
wood, 37
Crane fly wings, 155–156
Creep tests, 99–100,
  101f. *See also* Viscoelastic solids;
  transient tests
viscoelastic material, 155
Crossed-fiber helical array, 133–134, 134f
  helical fiber angles, 134–135, 135f
helically reinforced hydrostat, volume,
  length, and fiber angle, 134–135,
  136f
helically wound hydrostats, 134

Crossed-fiber helical array (*Continued*)
  helical reinforcing array, 134—135
Critical/Griffith crack length, 34—35
Currey, John D., 140
Cyclic/oscillating motion, 9

D

Daffodils, 47
d'Alembert's paradox, 55—56
Daniel, Thomas L., 4—5, 9, 87, 93—94
Defense Advanced Research Projects Agency
      (DARPA), 147
Deformation, 16. *See also* Strain
  microscopic deformation, 16—17
  plastic deformation, 22
  rubber band, 16—17
Denny, Mark W., 4—5, 66—67, 115—117
Dial, Kenneth, 4—5
Digital particle image velocimetry, 155
Drag
  acceleration reduction, 56
  as an asset, 62—63
  arbitrary free stream velocity, 54—55
  d'Alembert's paradox, 55—56
  defined, 54
  deflection, 54—55
  drag coefficient, 56—57
  drag equation, 56
  drag reduction, swimmers and flyers,
      60—62
    blunt/nonstreamlined reference value,
      60—61
    boxfish fins, 60
    California sea lions, 60—61
    diving beetle legs, 60
    drag coefficients, 60—61
    ducks and geese, 61—62
    emperor penguins, 60—61
    flat plate drag, 60—62
    frog tadpoles, 60—61
    "gliding" drag, 60
    indirect methods, 61—62
    isopods, 62
    locust flight, 62
    mackerel, 60—61
    mass and deceleration, 60
    Newton's second law, 60
    parasite drag, 60—61
    pigeons, 61—62
    power, 60
    primary decelerating force, 60
    rainbow trout, 61
      Reynolds numbers, 60—62
      sea lion flipper, 60
      streamlined reference value, 60—61
      undulatory swimming, 61
      vultures, 61—62
  dynamic pressure, 56
  fluid deceleration, 56
  fluid velocity, 54—55
  free-stream speed, 54—55
  frontal area, 56—57
  ideal fluid, 55—56, 55f
  induced drag. *See* Induced drag
  momentum, 56
  Newton's analysis, 55—56
  nondimensional index, 56
  nonstreamlined ("bluff") objects, 56—57
  no-slip condition, 54—55, 54b
  planform area, 56—57
  pressure drag, 55—57, 59—60, 59f
  real fluid, 55—56, 55f
  rear stagnation point, 56
  resistance, 54
  separation points, 56
  skin friction, 60
  solid obstruction, 54—55
  stagnation point, 55—56
  streamlined objects, 56—57
  unsteady drag forces, 93—94
  velocity gradient, 54—55
  viscous drag, 54—55, 57
  wing/lift-producing object, 56—57
Drag equation, 56
Drag coefficient, 56—57, 60—61
Ductile materials, 27—28
Ductility, 36
Dudley, Robert, 4—5
Dynamic pressure, 63
Dynamic tests for
      viscoelasticity. *See* Viscoelastic
      solids, dynamic testing
Dynamic viscosity, 53. *See also* Viscosity

E

Echinoderms
  sand dollars, 37
  sea urchins, 37
  starfish, 37
Eiffel, Gustav, 74—75
Elastic storage modulus, 105
Elastin, 24—25, 25b, 105
Ellington, Charles P., 4—5, 80—81
Emergency "tail flip" escape behavior, 87

Energy
  chemical bond, 6
  definition, 6
  heat energy, 6
  internal energy, 6
  kinetic energy (KE), 6, 63, 127—128, 127f
  law of conservation of, 3, 6
  muscle contraction, 142
  potential energy (PE), 6, 63, 68, 127—128, 127f
  storage mechanisms, 28—29
  storage/toughness, 23
  strain energy, 26, 34
    work of extension, 20—21, 21f
  stress—strain curves, 20
  thermal energy, 63
  work relationship. *See* Work
Ennos, A. Roland, 4—5, 47, 137
Enzymes, 114
Equations of motion, 9
Euler buckling, 42—43, 43f, 47
Euplectella aspergillum, 38

F
Fabrication methods, 152—153
Feather shaft, 45—46
Fexural stiffness, 40, 42—43
Fiberglass, 36, 38
Finite element analyses, 5
Fishlike robots, 149
Fish swim bladders walls, 3—4
Flapping
  active upstroke, 79, 80f
  anatomical lower surface functions, 80
  bound vortex, 80—81
  downstroke, 79, 79f
  fast-flying birds, 79
  flapping flyers, 78b
  flapping pattern, 79
  forward component, 79
  helicopter rotor, 77—79
  inactive upstroke, 79, 80f
  insect wings, 80—81
  leading edge vortex, 80—81
  medium-sized flying animals, 80
  "passive" upstroke, 79
  resultant force, 77, 80f
  robotic flapping model, 80—81
  thrust production, 77
  vertical component, 79
  wing's chord, 79

  wingbeat patterns, 79, 80f
Flapping flippers, 148
Flapping insect wings, 155—156
Flapping wing aerodynamics, 77—81, 78b, 156
  stroke cycle, 71—72
Flapping-wing flying machines, 147
Flight biomechanics, 4—5
  wingbeat cycle, 9
  wing hinges, 28
  wings and size
    clap-fling process, 84, 84f
    leading edge vortex, 83
    low-Reynolds number, 83
    quasisteady approximation, 83
    trailing edges, 84
    Wagner effect, 84
Flight mechanics, 156
Flow patterns, 155
Fluid biomechanics, 4—5
  advantages and disadvantages, 52
  Bernoulli's equation, 51, 63—64
  biological wings, 69
  circulatory systems, 52—53
  continuity principle, 52—53
  density, 52—53
  dippers, 69
  drag
    defined, 54
    drag coefficient, 56—57
    nonstreamlined ("bluff") objects, 56—57
    pressure drag, 55—56
    streamlined objects, 56—57
    viscous drag, 54—55
  drag reduction, swimmers and flyers, 60—62
  Eulerian perspective, 52
  flapping. *See* Flapping
  flow tanks, 52
  flying animals, 51
  gases, 51—52
  gliding. *See* Gliding
  hydrofoils, 81—82, 82f
  incompressible gases, 51—52, 51b
  induced drag. *See* Induced drag
  internal flows. *See* Internal flows
  internal fluid transport systems, 52—53
  Lagrangian perspective, 52
  lift, definition, 69
  lift mechanism. *See* Lift mechanism
  liquids, 51—52
  Navier—Stokes Equations, 64—65

Fluid biomechanics (*Continued*)
  preserved fish carcass, 52
  puffins, 69
  rate of deformation, 51—52
  rate of shearing, 51—52
  respiratory systems, 52—53
  Reynolds number. *See* Reynolds number
  sea lion flippers, 69
  swimming, 51. *See also* Swimming
  tuna tails, 69
  unsteady flows. *See* Unsteady flows
  velocity gradients. *See* Velocity gradients
  viscosity, 53—54, 53f
  volume flow rate, 52—53
  wind tunnels, 52
  wings and size, 82—84. *See also* Flight
    biomechanics; wings and size
Fluid flow patterns, 51b
Fluid mechanics, 3—4, 9
Fluid velocity, 58b
Force, 12
  definition, 8
  drag. *See* Drag
  force generation in climbing plants, 5
  lift force, 69
  muscle force, 123
  tension, 32, 34, 34f
  weight. *See* Weight
Fracture mechanics
  bent steel plate, 33—34
  brittle material scratch, 33—34
  catastrophic failure, 34
  chemical properties, 32
  crack propagation, 33—34
  cracks and flaws, 33—34
  critical/Griffith crack length, 34
  failure modes, 32
  flaws/voids, 32
  force trajectories, 32, 33f
  Hookean materials, 34
  isotropic material, 32—33
  load-bearing structures, 32
  localized stress, 32
  narrow cracks, 32—33
  plexiglas plates, 33—34
  rough/ragged fracture surface, 35
  semicircular notches/circular voids, 32—33
  smooth fracture surfaces, 35
  solid object, 35
  strain energy, 34
  stress, 32—33
  tensile strength, 32
    tension, 32, 34, 34f
    tough materials, 34—35
  Fractures in tree branches, 5
  Freely movable articulations, 113
  Frost, H.M., 1
  Froude efficiency, 89, 147
  Froude number, 131—132
  Functional systems, 121
  Fung, Y.C., 3

G

Gaits. *See also* Legged locomotion
  bipedal walk and run, 130b
  bounding, 130b
  canter, 130b
  compass, 127—128
  galloping, 130—131
  hopping, 129—130
  inverted-pendulum, 127—128
  pace, 130b
  running, 126—129
  tripod, 132
  trotting, 130—131
  walking, 126—128, 130—131
Galloping gaits, 130—131
Geckolike adhesive, 147—148
Geckos, 146—148
Gecko toe pads, 147—148
Glass, 152
  fibers, 36, 38
  stress—strain curves, 20
Gliding
  aerodynamic control, 78b
  bound vortex, 80—81
  equilibrium glide, 78b
  fixed-wing airplanes, 77—79
  glide ratio, 78b
  gliding geometry, 77, 78f
  leading edge vortex, 80—81
  resultant force, 77, 78f
  ridge lift, 77
  soaring, 77
  thermals and orographic lift, 77
Glycoproteins, 114
Greenstick fracture, 45, 45f
Goldman, Daniel I., 117—119
Gordon, James E., 32, 41,
    47—49
Griffith, Alan A., 33—34
Griffith crack length. *See* Critical/
    Griffith crack length
Griffith crack propagation, 35, 43

H

Hagen–Poiseuille equation, 91
Hawk moth wings, 155–156
Helical proteins, 25b
Hemoglobin molecules, 152
Hexapedal robots, 148
High-speed photography and video, 155–156
Hill, A.V., 141–142
Hookean elasticity, 21–22
Hookean materials, 34
Hooke, Robert, 3
Hooke's law, 3
    definition, 21–22
Hopping gait, 129–130
House flies, 147–148
House fly flight mechanics, 147
House fly wings, 155–156
Hovering insects, 156
Human-made ceramics, 152
Human-made structures, 152–153
Humanoid robots, 148–149
Hydrofoils, 81–82, 82f
Hydrostatic skeletons, 132–138, 153
    antagonist, 136–137
    biological hydrostats, 136–137
    bone, 132
    circumferential muscles, 137
    collagen fibers, 137
    continuous partial contraction, 136–137
    earthworms, 137
    fiber-reinforced hydrostats
        bending, 133–134
        buckling/kinking, 133–134
        cellulose, 133–134
        "cheap" tensile fibers, 133–134
        circular cross section, 134–135
        collagen, 133–134
        compression, 133–134
        crossed-fiber helical array, 133–134,
            134f
        extension, 133–134
        flaccid region, 134–135
        helical fiber angles, 134–135, 135f
        helically reinforced hydrostat, volume,
            length, and fiber angle, 134–135,
            136f
        helically wound hydrostats, 134
        helical reinforcing array, 134–135
        hoop-shaped fibers, 133–134
        noncircular cross section, 134–135
        orthogonally reinforced hydrostats, 134
        orthogonal pattern, 133–134, 134f

tension-resisting fibers, 133–134
fiber-reinforced outer body covering/
    cuticle, 136–137
flatworms, 137
helical reinforcing fibers, 137
herbaceous plants, 137
longitudinal and circumferential muscle,
    137
mantle (squid outer body wall), 137
muscles and stresses, 133
muscular hydrostats, 137–138
nematodes, 136–137
ribbon worms (Nemertea), 137
sea anemone mesoglea, 132
Hydroxyapatite crystals, 37–38

I

Ideal fluids, 55–56, 55f, 64
    inviscid (ideal) fluid theory, 93–94
Incompressible flow, 51b
Incompressible gases, 51–52, 51b
Induced drag
    aerodynamic efficiency, 76
    airplane (typical), 74
    aspect ratio, 75–76, 76f
    bird (typical), 74
    factors, 75
    gliding birds, 76–77
    lift production cost, 74
    lift-to-drag ratio (L/D), 74
    maximum lift and minimum drag, 74–75
    morphing wings, 77
    nonrectangular wings, 75
    polar plot, 74–75, 75f
    slotted tips, 76–77
    sparrow, 76
    swallows, 76
    thrust production, 74
    trees and underbrush, 76
    vulture, 76–77
    wing geometry, 77
Inertial drag, 56. *See also* Pressure drag
Insect cuticle, 152
    abdominal plates, 107–108
    arthropod cuticle, 17
    creep and stress-relaxation, 108
    elytra, 108
    exoskeleton, 107–108
    flexible cuticle, 23–24
    locust mandibles, 107–108
    modulus of elasticity, 108, 109f
    postfeeding cuticle, 108

Insect cuticle (*Continued*)
  stress-softening behavior, 108, 109f
  tanned, 28
Insect flight mechanics, 3—5, 155—156
  computer modeling, 148
  wing structure, 46
Insect joints, 23—24
Insect mouthparts, 3—4
Internal flows
  average flow speed, 91—92
  circular channel, 92
  correction factors, 92—93
  elephant respiratory passages, 93
  flow speed, 91—92
  Hagen—Poiseuille equation, 91
  hydraulic diameter, 92—93
  laminar flow, 90—91, 91f
  material economy and mechanical strength,
      90—91
  maximum flow speed, 91—92
  no-slip condition, 90—91
  in plants and animals, 90—91
  plug flow, 90—91
  resistance, 91, 92b
  Reynolds numbers, 90—91
  sensitivity, 92—93
  transport tubes, 92—93
  turbulent flow, 93
  velocity gradients, 90—91
  volume flow rate, 91
Internal fluid transport systems, 52—53
Intracellular movement mechanisms, 3
Inverted-pendulum mechanism, 127—128
Inviscid (ideal) fluid theory,
      93—94. *See also* Ideal fluids
Isotropic materials, 28, 32—33

J
Jellyfish mesoglea, 17
Jensen, Martin, 3—4, 83
Jet propulsion, 85, 89
Jetting effect, 112
Jointed limb, 121
  components, 121
J-shaped stress—strain curve, 23—24, 107
Jumbo jets, 154
Jump heights, maximum
  air resistance, 142—143
  of animals, 141—142, 141t
  apodeme (tendon-like cuticular strut), 142
  catch mechanism, 142
  elastic release, 142

  energy expenditure, 143
  energy per unit mass, 141
  fleas, 141—142
  grasshoppers, 141—142
  intrinsic muscle speed, 142
  jumping efficiency, 143
  kangaroo rats, 142
  "knee" articulation, 142
  resilin spring, 142
  Reynolds numbers, 142
  spring-and-catch arrangement, 142
  tree frogs, 141—142
  vertical velocity, 143
Jumping systems, 28

K
Kaye effect, 112
Kelvin—Voigt model, 101—102, 102f
Keratin, hooves, 108
Keratinous structures, 155
Kier, William M., 4—5, 138
Kinematics, 155—156
  bipedal running, 127—128, 127f
  inverted pendulum walking, 127—128,
      127f
  kinematic viscosity, 53
  solid body, 9
Koehl, Mimi, 4—5

L
Laminar flow, 90—91, 91f
Laplace equations, 133
Laser-based technique, 155
Laser speckle interferometry, 155
Leading-edge vortex, 80—81, 83—84, 96, 151,
      155—156
Leaping effect, 112
Legged locomotion, 148
Legs
  arthropods, 121
  articulation. *See* Articulation, rigid
      skeletons
  locomotion. *See* Legged locomotion
  muscle biomechanics and
      scaling. *See* Muscle biomechanics
      and scaling
  vertebrates, 121
Legged locomotion
  Achilles tendon, 129—130
  aerial phase, 126, 129
  bipedal, 126

bipedal hopping, 130b
bipedal runner, 129
centripetal acceleration, 128–129
deceleration, 128–129
duty factor, 126, 130–131
elastic energy, 130–131
energy conservation, 129
four-legged/quadrupedal animals,
    130–131
Froude number, 131–132
galloping gaits, 130–131
hexapedal, 126
inverted-pendulum mechanism, 127–128
legged locomotion, 126
medium-sized and large quadrupeds,
    130–131
metabolic costs and speed, 131–132, 132f
multilegged locomotion, 148
passive compression springs, 129–130
pole-vaulter's pole, 127–128
quadrupedal, 126
quadrupedal walk, trot and run, 130b
quadrupeds, 130b
ramp-walking toy, passive dynamic walker,
    127–128, 128f
stance leg, 129
stance/support phase, 126
stiff-legged walk, 127–128
stride, 126
stride lengths, 126–128, 130b
swing/recovery phase, 126
symmetrical gait, 130–131
three-legged stool, 132
walking racers, 128–129
Lift, 69
    definition, 69
Lift mechanism
    airfoil terminology, 70, 70f
    angle of attack, 71
    bound vortex, 71, 71f
    camber, 70
    chord, 70
    circulation, 71–72
    fluid theory, 71–72
    lift coefficient, 72
    lift force, 69
    lift modification
        angle of attack, 73, 73f
        Bernoulli's equation, 74
        camber, 73
        downwash, 72–73
        flying animals, 74

gliding birds, 73–74
lift enhancement, 73
lift production, 72–73
morphing wings, 73–74
Newton's third law, 72–73
nutshell, 73
shape/orientation, 73
speed's effect, 74
stall, 73
trailing edge, 72–73
vortex system, 72–73
lift production, 70–71
net vortex, 71
nondimensional index, 72
planform area, 72
sharp trailing edge, 71
starting vortex, 72f
stopping vortex, 71–72
stroke cycle, 71–72
tip vortices, 71–72, 72f
wind tunnel tests, 69–70
wing cross section/airfoil, 69–70, 69f
Lift-to-drag ratio (L/D), 74
Linear momentum, 8
Load carrying, 139
Local buckling, 43–44
Locust flight mechanics, 3–4
Locomotion. See Flapping, Gliding, Legged
    locomotion, Swimming
Lomakin, Joseph, 108
Longitudinal stress, 133
Looping effect, 112
Lumber, dried ("seasoned"), 105

M
Macroscopic organisms, 2–3
Manometric pressure, 63
Manometric pressure equation, 64
Masonry, 152
Mass and deceleration, 60, 93–95
Maxwell model, 101–102, 102f
Medical biomechanics. See Clinical
    biomechanics
Medusan jellyfish, 89
Membrane potentials, molecular biology, 3
Mesoglea, 110–111, 132, 152
Metals, 152
Microaerial vehicles (MAVs), 147, 156
Micromechanical Flying Insect of Ron
    Fearing's lab, 147
Modulus of elasticity,
    19–20. See also Young's modulus

Modulus of elasticity (*Continued*)
  tangent moduli, 30
Molecular mechanisms, 3
Mollusk shells, 37–38, 48–49, 152
Moment arm, 41–42, 41f
Moment of force, 41–42
Moment of inertia, 40
Momentum, 8, 56
  elastic collision, 8
  principles, 63
  rotational inertia, 8
  torques/moments, 8
Mucus, 121
  air passages, 114–115
  aquatic animals, 114–115
  burrowing marine annelid, 114–115
  clams, 114–115
  digestive system, 114–115
  echiurid worms, 114–115
  gliding locomotion, 116–117
  glycoproteins, 114–115
  hagfish, 114–115
  interwaves, 116–117
  jellyfish, 114–115
  mucins, 114–115
  Northwest Pacific banana slugs, 116–117
  particle-capturing structures, 114–115
  pedal mucus, mechanical properties,
      116–117
  pelagic tunicates, 114–115
  respiratory system, 114–115
  sea anemones, 114–115
  sessile gastropods, 114–115
  sessile tunicates, 114–115
  slug mucus properties, 116–117
  snail pedal mucus. *See* Snail pedal mucus
  snail's foot, 116–117, 116f
  yield stress and strain, 116–117
Multilegged locomotion, 148
Muscle, 4–5, 15
  biomechanics
    actin and myosin, 121–122
    antagonistic muscle, 122
    ant's muscle force, 122–123
    champion athletes, 124
    contraction speed, 123
    elastic hinge ligament, 122
    force (tension) and speed, 122
    force production, 123
    force vs. speed, 123–124, 123f
    internal combustion engines, 123–124
    intrinsic speed, 123

  invertebrates, 122
  maximum specific power output, 124
  muscle tissue functions, 122
  myosin, 123
  shell-closing muscle, 122
  shortening speed, 123–124
  skeletal muscular system, 121–122
  sliding filament arrangement, 122
  sliding-filament mechanism, 121–122
  squid tentacles, 122
  tensile strength, 122
  terrestrial mammals, 123
  zero speed, 123–124
  scaling, 7f, 121–123
Muscle and soft tissue mechanics, 4–5
Mussels, 147–148
Mutations, 154

N
Nachtigall, Werner, 4
Natural rubbers, 25b
Natural selection, 151–152
Navier–Stokes equations, 9, 64–65, 156
Neutral density particles, 155
Newtonian physics
  explicit vector algebra, 5
  force, 12
  light intensity, 13
  mass, 10–12, 11f
  power, 9–10
  pressure, 10
  scalars, 5
  solid body kinematics, 9
  stress, 10
  temperature, 13
  unsteady motion, 9
  vectors, 5
  voltage, 13
  weight, 10–11, 11f
  work–energy relationship, 5–6
Newton's analysis, 55–56
Newton's first law, 6b
Newton's Laws of Motion
  first law, 6b
  second law. *See* Newton's second law
  third law, 6b, 16, 72–73
Newton's second law, 6b, 7f, 60
  acceleration, 7
  force, 8
  mass, 8
  speed, 7
  velocity, 7

Newton's third law, 6b, 16, 72—73
Niklas, Karl J., 144—145
Nonconservative work. *See* Work,
    nonconservative
Non-Hookean biological materials, 23—24,
    24f
Nonlinearity
    cast iron, 23
    collagen, 23—24
    compression-resistant materials, 23—24
    elastin, 24—25
    energy storage/toughness, 23
    flexible cuticle, 23—24
    glass, 23
    helical polypeptide chains, 23—24
    hypothetical biological material, 26, 26f
    hysteresis, 26
    J-shaped stress—strain curve, 23—24
    linearly elastic range, 23
    loading stress—strain curve, 26, 26f
    mammalian bone, 23—24
    non-Hookean biological materials, 23—24,
        24f
    nonlinear stress—strain curves, 23—24
    plastic range, 23
    pliant/tensile biomaterials, 26
    protein rubbers, 24—25
    relaxing curve, 26
    resilience, 24—25
    resilin, 24—25
    reversible molecular reconfigurations, 26
    rubber elasticity, 25b
    shock absorber/safety relief, 23
    skin and arterial walls, 23—24
    spider silk, 23—24
    spider web, 26
    steel/aluminum alloy, 23
    strain energy, 26
    tendons, 23—24
    tensile structures, 23—24
    unloading stress—strain curve, 26, 26f
    wet clay, 23
    yield stress, 23
Non-Newtonian liquids
    anecdotal observations, 112
    biological liquids, 114
    blood. *See* Blood
    cornstarch slurry, 111—112
    Kaye effect, 112
    polymer suspension, 112
    rheopectic, 112
    shear rate, 111

shear thickening, 111
shear thinning, 111
synovial fluid. *See* Synovial fluid
thixotropic, 112
time-dependent behavior, 112
viscoelastic solid, 112
viscosity, 111
Norberg, Ulla, 4
No-slip condition, 54—55, 54b,
    90—91

O

Ogston, A.G., 113—114
Organismal biomechanics, 2—3
    fluid—solid interactions and properties, 4
    Hooke's law, 3
    quantitative, engineering-inspired
        approach, 3—4
    Young's modulus, 3
Organismal *vs.* technological design
    biomimetics/biomimicry, 151
    blimps, 153
    connections and joints, 153
    construction process, 154
    delta-winged aircraft, 151
    human-built vehicles, 154
    leading-edge vortex, 151
    lever systems, 153—154
    "live" hinges, 153
    locomotion mechanism, 154
    materials, 152
    mechanics of nature *vs.* mechanics of
        engineers, 152
    natural selection, refining effects, 151—152
    pressurized hydrostats, 153
    pressurized structural components, 153
    research and methods, 155—156
    rotating wheel/axle system, 154
    shape, 152—153
    tasks and functions, 151—152
    technology transfer, 151
    terrestrial locomotion, 154
    venus fly traps, 153
    vertebrate/arthropod skeletons, 153—154
    vertebrate skeletal joints, 153
Osmotic pressure, 133
Oysters, 147—148

P

Papanicolaou, M.N., 155
Parle, Eoin, 44

Particle density, 155
Pathline, 58b
Pennycuick, Colin, 4
Pipes, 152—153
Placet, V., 105
Plant vascular transport systems, 64
Pliant/rigid biomaterials
  bone, 28
  composites, 28
  mollusk shell, 28
  stiff tensile fibers, 28
  tanned insect cuticle, 28
  tooth enamel, 28
  wood, 28
Poiseuille, Jean, 3
Poiseuille's equation, 3
Polar second moment of area, 42
Potential energy, 63
Power, 9—10
Prandtl, Ludwig, 65, 68
Pressure drag, 55—57,
    59—60, 59f
Protein gel matrix, 37
Protein/mucopolysaccharide, 28
Protein structure and function, 3
Proteoglycans, 106—107
Pterosaur wing mechanics, 5

Q
Quadrupedal robots, 148
Quasisteady approximation/analysis, 9, 95

R
Radio-controlled/autonomous flying
    machines, 147
Rat uterus, 3—4
Rayner, Jeremy, 4
Real fluid, 55—56, 55f
Relative viscosity, 53
Resilience, 24—25
Resilin, 17, 25b, 26
  flea leg, 28—29
  novel material, 29
  resilience, 24—25
  single component, 28
  wing hinge, 28—29
Resistive force theory (RFT) analysis,
    117—118, 118f
Reverse engineering and microfabrication,
    147
Reynolds numbers, 51, 90—91

aquatic diving beetles, 60
average flow velocity, 59
biological consequences, 57—58
biomechanical process, 57—58
chaotic swirls and eddies, 58—59
cylindrical strut, 59—60
discrete states, 58—59
dye stream, 58—59
flow direction, 59—60
laminar/turbulent flow, 58—59
molecular level process, 59
myriad tiny curls and eddies, 58—59
"normal" orientation, 59—60
physiological process, 57—58
pressure drag, 57, 59—60, 59f
reference length, 57
skin friction drag, 60
streamlines, 58—59, 58b
temperature and wall surface roughness,
    58—59
velocity gradients, 57—58
viscous drag, 57
viscous forces, 57
Reynolds, Osborne, 58—59, 108
"Riblet" technology, 149
RoboBee of Robert Wood's lab, 147
Robots
  fishlike, 149
  humanoid, 148—149
  legged, 148—149
  mechanical and control systems, 148
  quadrupedal and hexapedal, 148
  robotic flapping model, 80—81
  swimming, 148
Rubber elasticity, 24—25, 25b
  amorphous, 25b
Running gait, 126—129

S
Saliva, 114
Scallops, springy hinge ligament, 3—4
Scincus scincus (sandfish lizard), 117—118
Sea anemone mesoglea,
    23—24. *See also* Mesoglea
  body wall, 3—4
  collagen fibers, 110—111
  Metridium senile, 110—111, 111f
  retardation time, 110—111
  stress-relaxation test, 110—111
Secomb, T.W., 113
Second moment of area, 40—42, 40f
Setae, 147—148

Shadwick, Robert E., 106
Shafts, 152—153
Shear, 16, 16f
   modulus, 22—23
   stress, 22—23, 23f, 53
Shear thinning
   house paints, 111
   quick sand, 111
   whole blood, 111
Shells, 121
   crushing and piercing resistance, 44—45
   domelike meridians, 44—45
   eggshells, 44—45
   hemispherical domes, 44—45
   meridional compression, 44—45
   pressurized shells and domes, 45
   sea urchin skeletons, 44—45
   skulls, 44—45
   squeezing loads, 44—45
Sinking speed, 62—63
Sisal leaf fibers, 3—4
Size consequences
   aerodynamic regimes, 143—144
   ant muscles, 138—139
   apical meristem, 144—145
   biomimicry. *See* Biomimicry
   California redwoods, 145—146
   cellulose fibers, 144
   conifers, 145
   cylindrical stem, 145—146
   elastica theory analysis, 145
   endurance, 144
   epidermal cells, 144
   flowering plant, 145
   fossils, 143—144
   garlic stalks, 145
   geometric similarity, 144
   Giant sequoias, 145—146
   heartwood, 145
   herbaceous plants, 145
   herbaceous (nonwoody)
      plant stem, 144
   horsetails, 145
   hydrostatic tissues, 146
   insects, 138—139
   land plants, 144—145
   lycopods, 145
   mammals, 138—139
   maneuverability, 144
   maximum jump heights. *See* Jump heights,
      maximum
   multicellular animals, 138—139

osmotic process, 144
parenchyma cells, 144
pterosaurs, 143
reproductive stalks, mosses, 144
Reynolds numbers, 143—144
Rhamphorhynchus, 143—144
sapwood, 145
sclerenchyma, 145
secondary growth, 145
secondary xylem, 145—146
soaring ability, 144
Solnhofen limestone, 143—144
Southeast Asian rainforests, 145—146
surface to volume ratio (S/V). *See* Surface
   to volume ratio (S/V)
takeoff speed and angle, 144
tree seedlings, 146
tropical rainforest canopies, 145—146
turgor pressure, 144
vascular plants, 144—145
water-resistant protein lignin, 145
wingspan, 143—144
Skeletons. *See also* Hydrostatic skeletons
   bird wing bones, 45—46
   exoskeleton, 29—30, 107—108
   hollow vertebrate bones, 45
   leg bones, 47
   mammalian bone, 23—24
   mammalian skeletons, 139—140
   mammal leg bone, 47
   skeletal proportions, 139
   skeletal support systems, 139—140
   spongy bone, 29
Skin, 15
   elastic component, 107
   J-shaped stress—strain curve, 107
   noninvasive dynamic mechanism, 107
Skin friction, 54—55
   drag, 60
Snail pedal mucus
   Bingham plastics, 115
   elastic strength, 115
   flow stress, 115—116
   foot, 115
   gastropod mollusks, 115
   shear modulus, 115
   strain rate, 115
   viscous liquid, 115—116
   yield strength, 115
Soaring, 77, 144
Soft (hydrostatic) skeletons. *See* Hydrostatic
   skeletons

Soft tissue
  displacements, 155
  mechanical properties, 4–5
Solid materials
  aluminum alloys, 17
  anisotropic, 28
  aquatic crustaceans, 29–30
  arthropod cuticle, 17
  automatic movement control, 28–29
  biological solid materials, 15
  bone, 15, 17
  calcium carbonate, 29–30
  Cauchy strain, 18
  compression, 16, 16f
  deformation, 16
    microscopic deformation, 16–17
    plastic deformation, 22
    rubber band, 16–17
  dried and green twigs, 28
  elasticity, solid's resistance, 15
  energy storage mechanisms, 28–29
  failure and prevention, 30–32
    crack growth inhibition. *See* Crack
      growth inhibition
    fracture mechanics. *See* Fracture
      mechanics
  fleas, 28–29
  flexible materials, 17
  floppy materials, 17
  force-extension data, 17–18, 18f
  gelatin/saltine crackers, 30
  isotropic, 28
  kangaroos, 28–29
  kiln-dried wood, 28
  masonry and unreinforced concrete, 17
  migratory locusts, 28–29
  mineral crystals, 29–30
  molecular arrangement, 16
  muscle, 15
  natural material, 16–17
  Newton's third law, 16
  nonlinearity. *See* Nonlinearity
  pliant/rigid biomaterials, 28
  polymer molecules, 30
  properties, 30, 31t
  protein–chitin composite exoskeleton,
    29–30
  reactive force, 16
  resilient biomaterials, 28–29
  resilin pads, 28–29
  rubber-like proteins, 29
  shape and size, 17–18
  shear, 16, 16f
  skin, 15
  sledgehammer blow, 28
  spongy bone, 29
  steel, 15, 17
  stiff steel wire, 29
  strength *vs.* toughness. *See* Strength *vs.*
    toughness
  stress, 17–18
  stress–strain curves. *See* Stress–strain
    curves
  stretchy materials, 17
  structures. *See* Structures
  synthetic material, 16–17
  tangent moduli, 30
  tendons, 17, 30
    rope/biological counterparts, 17
  tensile biomaterials, 28
  tension, 16, 16f
  tension-resisting engineered structures, 17
  time-dependent/viscoelastic behavior, 17
  tooth enamel/glass, 30
  toughness, 28
  true/natural strain, 19
  unstressed length/strain, 18
  viscoelastic properties, 15
  viscosity, fluid's resistance, 15
  wood, 15, 17
    and bone, 30
    fibers, 29–30
    and insect cuticle, 29–30, 30f
  Young's modulus, 28, 30
Sonomicrometry, 155
Spatula (on gecko feet), 147–148
Spedding, Geoffrey R., 58b
Spider silk
  elastic modulus, 110
  loss modulus, 109–110
  "major ampullate" silk, 109–110
  and snail mucus, 4–5
  storage modulus, 109–110
Sports biomechanics
  athletic performance, 2
  normal locomotion mechanics, 2
  prosthetic limbs, 2
Springs and dashpots models, 101–102, 102f
Standard linear solid model, 101–102, 102f
Stanier, J.E., 113–114
Static pressure, 63
Steel, 15, 17, 21–23, 21f, 26–27, 27f, 137
  ductility, 36
  fracturing and bending, 33–34

high strength/toughness, 30
steel springs, 28–29
spring steel rod, 47
Stiffness, 17. *See also* Young's modulus
  composites, 36
  fexural, 40, 42–43
  rubber stiffness, 25b
  stiff material, 19–20, 22–23
  tensile biomaterials, 28
  torsional, 42
Stoke's law, 62–63
Strain, 19
  Cauchy strain, 18
  energy, 26, 34
  engineering strain, 18–19, 21f, 100f
  natural strain, 19
  strain rate, 51–52, 103–104, 103f, 108,
    115–116
  unstressed length, 18
Streakline, 58b
Streamlines, 55f, 58–59, 58b, 63–64, 68
Streamlining, 56–57, 59–61, 59f
Strength, 20–21
  ultimate strength, 20
Strength *vs.* toughness
  aluminum alloys, 27–28
  brittle material, 26–27, 27f
  cast iron, 26–27, 27f
  dry twig, 26–27
  giant seaweed, 27–28
  green twig, 26–27
  mild steel, 26–27, 27f
  spider frame silk, 27–28
  tough material, 26–27, 27f
  Young's moduli, 26–27
Stress, 10
  beam stresses, 39–40
  circumferential, 133
  columns, 43
  definition, 17–18
  fracture, 32–33
  localized, 32
  longitudinal, 133
  mucus yield stress, 116–117
  muscle, 133
  shear, 22–23, 23f, 53
  tensile and compressive, 40
  viscoelastic, 99–100
  yield stress, 22–23
Stress–strain curves
  aluminum alloys, 21–22
  axes of, 19b

bone's plastic behavior, 22
brittle materials, 20–21
cartilage, 21–22
cast iron, 21–22, 21f
ceramics, 21–22
compliant, 19–20
cortical bone, 20–21
expansion joints, 20
extensibility, 20
force-extension test, 19
glass, 21–22, 21f
Hookean behavior, 21–22
Hookean elasticity, 21–22
human skin, 20
independent and dependent variables, 19b
insect exoskeletons, 20
internal force distribution, 19b
keratin, 21–22
linearly elastic behavior, 21–22
low-carbon steel, 21–22, 21f
postyield strain, 22
properties, 20
scale-independent form, 20
shear modulus, 22–23
shear stress, 22–23, 23f
spider silk, 20
steel, 21–22
stiff materials, 19–20
strain energy/work of extension, 20–21,
  21f
strength, 20–21
tensile stress–strain curves, 23
toughness, 20–21
ultimate strength, 20
vertebrate compact bone, 22
yield stress, 22
Young's modulus, 19–20, 22–23
Strouhal number, 9
Structures
  aerodynamic forces, 47–48
  aerodynamic load, 47
  anisotropic arrangement, 45
  arthropod legs, 45
  banana leaves petioles, 47
  beams, 38. *See also* Beams
  bending, 47
  bending and buckling, 38
  bending and torsion, 45–46
  bending compliance, 46–47
  bending/twisting damage avoidance, 45
  bicycle spokes, 47
  biological struts function, 47–48

Structures (*Continued*)
 bird eggshells, 49
 bird wing bones, 45–46
 bulkheads, 47
 calcium carbonate, 48–49
 cellulose fibers, 45
 columns, 38. *See also* Columns
 compression-resisting bulkheads, 47
 cross-sectional area, 38
 crystalline arrangements, 48–49
 daffodils, 47
 downwind side compression, 47–48
 elongated biological supports, 45
 engineering hierarchy, 38
 Euler buckling, 47
 feather shaft, 45–46
 flat aragonite crystals, 48–49
 flat plates, 46
 flattened pleats, 46
 flexural and torsional strength, 45
 gravitational compression, 47
 greenstick fracture, 45, 45f
 hollow vertebrate bones, 45
 horizontal tree branch, 45
 insect wings, 46
 leg bones, 47
 lignin, 45
 loading, types, 38
 longitudinal stiffeners, hollow cylinders,
   47, 48f
 mammal leg bone, 47
 mammal skulls and turtle shells, 49
 maple leaf petiole, 47
 mineralized layers, 48–49
 Mollusk shells, 48–49
 mother of pearl/nacre, 48–49
 muscles and hemolymph (blood), 45
 ovalization, 47
 pre-tension reduction, 47–48
 primary/pinion feather, 45–46
 protein matrix, 48–49
 quantitative mechanical analyses, 48–49
 shells, 38. *See also* Shells
 snails and clams, 48–49
 spiders, 45, 46f
 spring steel rod, 47
 stabilization, 48–49
 tensile strength, 45
 tension-resisting bulkheads, 47
 torsional compliance, 46–47
 transverse partitions, hollow cylinders, 47, 48f
 tree trunks, 47–48

 twist-to-bend ratio, 46
 violent tropical storms, 47
 wind-pollinated sedges, 47
 wood, 45, 47–48
Struts, 121
Surface to volume ratio (S/V)
 elephants, 139
 first-order approximations, 139
 gas exchange, 139
 geometric similarity/isometry, 139
 honeybees, 139
 hummingbirds, 139
 insects and spiders, leg dimensions, 140
 king crabs, 140
 limb bones, tiger, 139–140
 load carrying, 139
 localized impact loading, 140
 Maine lobsters, 140
 mammalian skeletons, 139–140
 opilionid harvestmen/daddy-longlegs, 140
 skeletal proportions, 139
 skeletal support systems, 139–140
 temperature regulating ability, 139
 terminal velocity, 139
 tyrannosaurs, 139
 Tyrannosaurus rex, 139
Swart, P., 82–83
Swartz, Sharon M., 4–5
Swimming, 51
 density and viscosity values, 84–85
 drag-based swimming
  aquatic arthropods, 85–86
  drag configuration, 85–86
  emergency escape behavior, 87
  hydrofoils, 86–87
  muskrat/freshwater turtle, 86–87
  recovery stroke, 85–86
  stroke pattern, 86–87
  swimming arthropods, folding bristles,
    85–86, 86f
 by jetting, 89
 lift-based swimming, 85
 locomotion mechanism, 84
 at low Reynolds numbers, 89–90
 modes, 85
 Reynolds numbers, 84–85
 robot, 148
 in sand, granular media locomotion
  Bingham plastic, 119
  dry sand, 117
  nematodes, 119
  polychaete worms, 119

resistance, 117
resistive force theory (RFT) analysis,
   117–118, 118f
   sandfish lizard, 117–118
   sand grains, 117
   segment axis and tangential force,
     117–118
   segment orientation angle, 117–118
   tangential resistance, 119
   undulatory swimming. *See* Undulatory
     swimming
Synovial fluid
   articulations (joints), 113
   bearing surfaces, 113–114
   dynamic testing, 113–114
   hyaluronic acid's molecular structure,
     113–114
   physical behavior, 113–114
   shear rate, 113–114
   shear thickening, 113–114
   shear thinning, 113–114
   static loading, 113–114
   synovial joints, 113
   vertebrate animals, 113

T

Tail beat mechanism, 81–82
Tendon, 17, 23–24, 30, 37, 99–100
   Achilles, 129–130
   rat tail, 3–4
Tensile failure and wall rupture, 133
Tensile materials, 17, 28
Tensile strength, 32
Tension, 16, 16f, 22–23
   elongation strain, 39–40
Terminal velocity, 62–63, 139
Test floppy/compliant materials, 155
Thermal energy, 63
Thom, A., 82–83
Thumb-sized turbine (jet) engines, 147
Thunniform, 87–88, 88f
Tiny ornithopters, 149
Tissue biomechanics, 4–5
Torque/moment, 41–42
Torsional stiffness, 42
Toughness, 20–21, 28, 35,
   37–38. *See also* Strength
    vs. toughness
   energy storage, 23
Transient tests for viscoelasticity. *See*
   Viscoelastic solids, dynamic testing
Tucker, Vance A., 61–62

U

Undulatory swimming
   bowfin, 88
   carangiform swimming, 88–89
   eels, 87
   elongate slender body theory, 87
   flow visualization, 88–89
   gymnotiform knifefishes, 88
   hydrodynamics of, 87–88
   ichthyological terminology, 87–88,
     88f
   leeches, 87
   polychaete worms, 87
   "reactive" model, 87
   sea snakes, 87
   thrust production, 87
Units
   calorie, 13
   gigapascals (GPa), 18
   megapascals (MPa), 18
   metric system, 13
   newton-meter (Nm), 41–42
   pascal, 18–20
   pound, 12
   pseudometric units, 13
   SI (International System of Units), 13
   slug, 12
   US customary system, 12
University of California Berkeley, 147
Unsteady flows
   acceleration reaction concept, 93–94
   added mass coefficient, 93–94
   additional force, 93–94
   in air, 95–96
   barnacles, 95
   drag forces, 93–94
   flow patterns, 93
   inviscid (ideal) fluid theory, 93–94
   mussels, 95
   negative acceleration/deceleration, 93–94
   positive acceleration, 93–94
   total force, 93–94
Unsteady swimming mechanics, 4–5

V

van der Waals attraction, 147–148
Velocity
   boundary layer flow, 66–67
   drag and, 54–55
   fluid, 58b
   Reynolds number and flow, 59
   terminal, 62–63, 139

Velocity (*Continued*)
vertical, maximum jump height, 143
Velocity gradients, 54—55, 57—58,
90—91. *See also* Boundary layers
fluid velocity, 65
no-slip condition, 65
solid surface, 65
Venus's flower basket, 38
Versluis, M., 112
Videler, John J., 87
Vincent, Julian F.V., 4, 101—102, 104, 108,
146
Virtual lever system, 138
Viscoelastic solids
arteries, 105—106, 106f
biological material, 99
bone, 105
cartilage, 106—107
dynamic testing
angular velocity, 104
elastic material, 103—104, 103f
elliptical stress—strain curve, 103—104
frequency/range of frequencies,
102—103
imaginary/viscous modulus, 104
linear viscoelasticity, 104
loss modulus, 104
nonlinearly viscoelastic materials, 104
phase angle, 104
real/elastic modulus, 104
sinusoidal loading, 102—103
sinusoidal motion, 104
sinusoidal strain, 103—104
storage modulus, 104
strain rate, 103—104
viscoelastic damping, 104
viscous liquid, 103—104, 103f
elastic material, 99
insect cuticle, 107—108, 109f
keratin, hooves, 108
sea anemone mesoglea, 110—111, 111f
skin, 107
spider silk, 109—110
springs and dashpots models, 101—102,
102f
test/rate of loading, 99
transient tests
constant stress, 99—100
creep test, 99—100, 101f
deformation, 99—100
dog knee ligament, stress—strain curves,
99, 100f

elastic modulus, 100—101
initial/instantaneous modulus, 100
mesoglea, 99—100
modulus of elasticity, 100, 101f
relaxation time, 100—101
retardation time, 100
sea anemones, 99—100
shear modulus, 100—101
stress-relaxation test, 100—101, 101f
tendon, 99—100
unstrained length, 99—100
viscosity, 100—101
wood, 105
Viscosity, 15, 53—54, 53f, 56
kinematic viscosity, 53
non-Newtonian, 111
pressure drag, 55—56
relative viscosity, 53
Reynolds number, 57—60
skin friction, 54—55
smallest animals, 62—63
viscous forces, 57
Vogel, Steven, 4, 27—28, 44—47, 49, 53—54,
56—57, 60—62, 64—65, 71, 86—87,
92—93, 105, 110—111, 134—135,
137—139, 141, 155
Volume flow rate, 58b
Von Kármán vortices, 95, 105
Vulcanization process, 25b

W
Wagner effect, 84, 96
Wainwright, Stephan A., 4, 17, 28, 49
Walking gait, 126—128
Wardle, C.S., 87
Webb, P.W., 94
Weight, 10—12, 11f
Weis-Fogh, Torkel, 3—4, 83—84, 84f, 96
Wingbeat cycle, 9
Wing hinges, 28
Wood, 15, 17, 29—30, 105
and bone, 30
components, 45
fibers, 29—30, 37
greenstick fractures, 45, 45f
green wood, 28
heartwood, 145
and insect cuticle, 29—30, 30f
kiln-dried wood, 28
sapwood, 145
secondary growth, 145
secondary xylem, 145—146

splitting wood, 28
tree height, 40
wood pre-stressed in trees, 47–48
Wood, Robert J., 108
Work, 63
   classical Newtonian mechanics, 5–6
   closed system, 6
   definition, 6
   law of conservation of energy, 6
   nonconservative, 6
   stress–strain curves, 20, 21f
   voltage energy, 6

Y

Young, Robert, 3
Young's modulus, 3, 26–28, 30, 105
   compression, 22–23
   insect cuticle, 107–108
   tangent moduli, 30
   tension, 22–23

Z

Zhang, T.N., 119

Printed in the United States
By Bookmasters